Working Watersheds

Bill Conlogue

WORKING WATERSHEDS

Water and Energy in the Lackawanna Valley

TEMPLE UNIVERSITY PRESS
Philadelphia • Rome • Tokyo

TEMPLE UNIVERSITY PRESS
Philadelphia, Pennsylvania 19122
tupress.temple.edu

Copyright © 2025 by Temple University—Of The Commonwealth System of Higher Education
All rights reserved
Published 2025

Library of Congress Cataloging-in-Publication Data

Names: Conlogue, William, author.
Title: Working watersheds : water and energy in the Lackawanna Valley / Bill Conlogue.
Description: Philadelphia : Temple University Press, 2025. | Includes bibliographical references and index. | Summary: "The Lackawanna Valley's history reveals how energy production and water circulation created positive feedback loops of technological and environmental change. To cope with climate change, which is water change, we need a change of mind; to survive in our warming world, we must look to history and literature as guides"— Provided by publisher.
Identifiers: LCCN 2024016398 (print) | LCCN 2024016399 (ebook) | ISBN 9781439926161 (cloth) | ISBN 9781439926178 (paperback) | ISBN 9781439926185 (pdf)
Subjects: LCSH: Water-supply—Pennsylvania—Lackawanna River Valley. | Energy industries—Pennsylvania—Lackawanna River Valley. | Climatic changes—Pennsylvania—Lackawanna River Valley. | Lackawanna River Valley (Pa.)—History. | Lackawanna River Valley (Pa.)—In literature. | Lackawanna River Valley (Pa.)—Environmental conditions.
Classification: LCC TD224.P4 C66 2025 (print) | LCC TD224.P4 (ebook) | DDC 628.1/0974836—dc23/eng/20240724
LC record available at https://lccn.loc.gov/2024016398
LC ebook record available at https://lccn.loc.gov/2024016399

♾ The paper used in this publication meets the requirements of the American National Standard for Information Sciences—Permanence of Paper for Printed Library Materials, ANSI Z39.48-1992

Printed in the United States of America

9 8 7 6 5 4 3 2 1

The best you can do is break even.
You'll never break even.
You end up broke, at even.

—LAWS OF THERMODYNAMICS

The troll said, "The secret is, '*Watch with both eyes!*'"

—JOHN GARDNER, *ON MORAL FICTION*

More than most places, Pennsylvania is what lies beneath.

—JENNIFER HAIGH, *HEAT AND LIGHT*

Contents

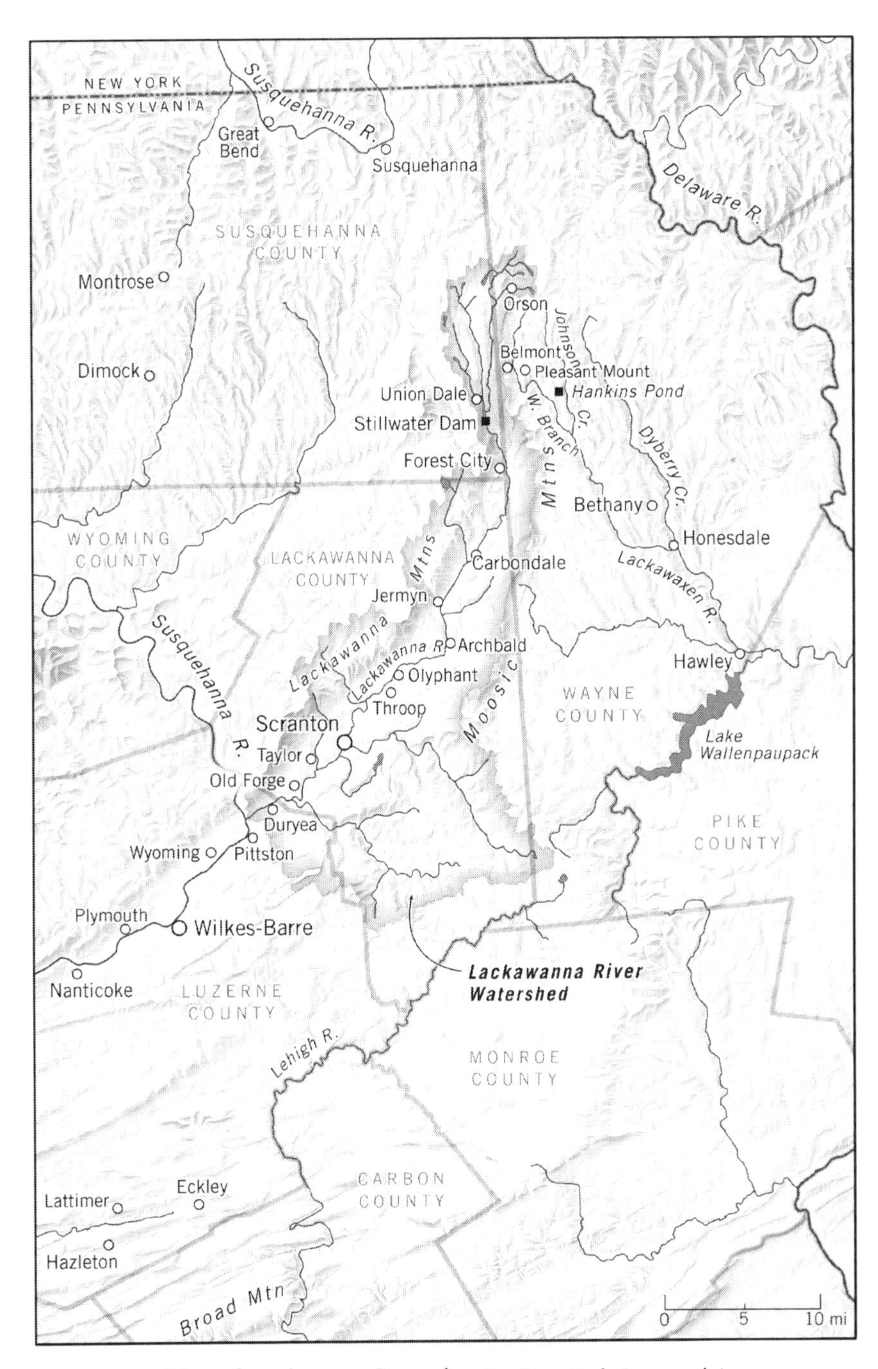

Map of northeastern Pennsylvania. *(Erin Greb Cartography)*

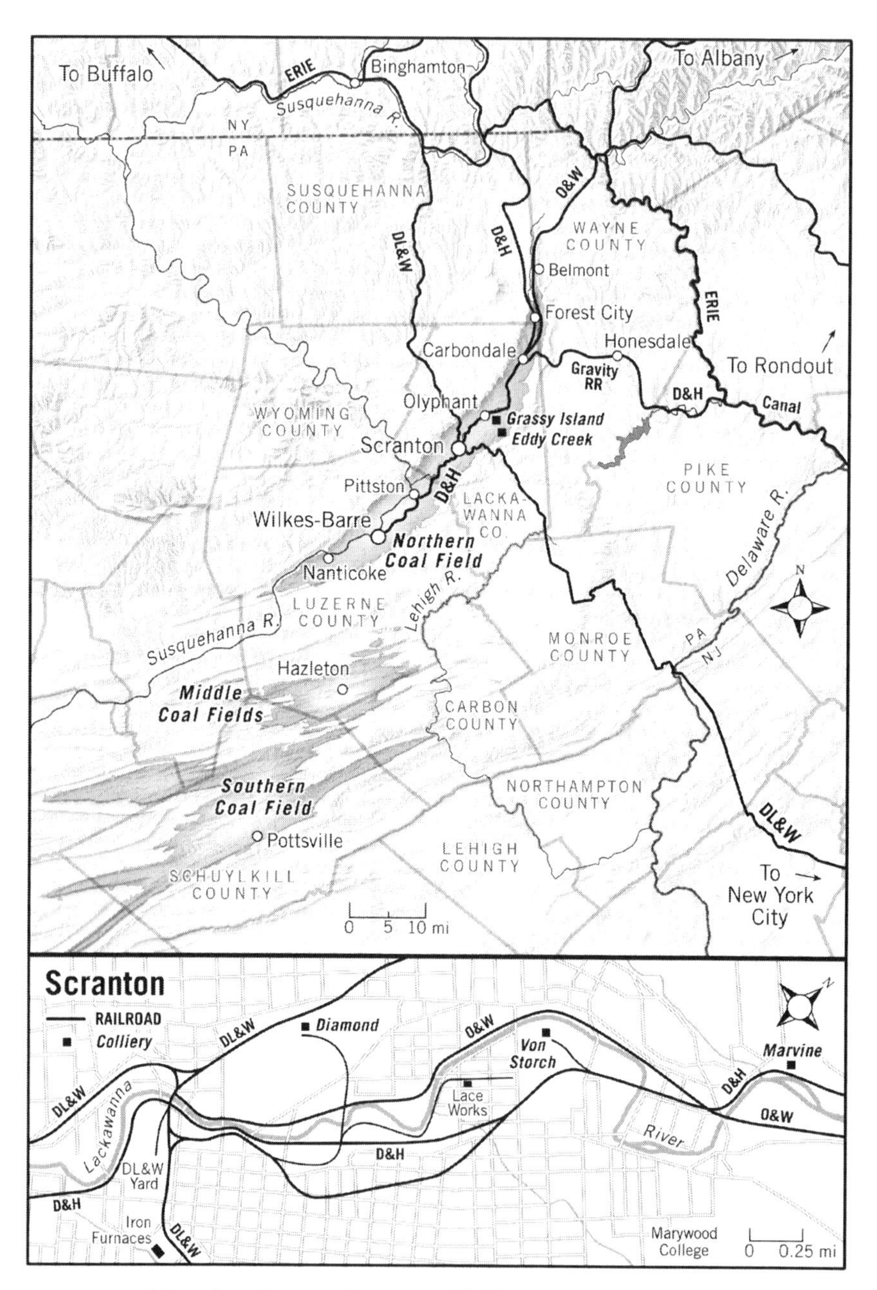

Map of northern anthracite coalfield. *(Erin Greb Cartography)*

Prologue

I wrote *Working Watersheds* as a historical *and* literary narrative. The book includes not only historical storytelling and literary analysis but also first-person anecdotes, maps, a timeline, and a glossary of coal-mining terms. I braid these together to capture the interconnectedness of reality. *Working Watersheds* is a complex text, one that I invite you to weave with me.

The introduction teaches people how to read the book: expect disruptions, connect form and content, and link local and global. The first chapter orients readers to deep time and how others have written about the Anthracite Region. Telling an origin story of anthracite mining, Chapter 2 unfolds chronologically, in the way of most conventional histories. Upending this traditional approach, Chapter 3, which explores how mining dealt with water, ranges across time and space, mimicking water cycles. Chapter 4 disrupts attention to coal mining by studying water in textile making, once the region's second most important industry. At the center of the book, Chapter 5 discusses water pollution and begins a set of chapters about natural and human responses to altered water cycles: flood, drought, conflict, and electricity generation. Chapter 10 interrupts this cascade of historical case studies by exploring the present natural gas boom, which returns the region to its history of fossil fuel production. A conclusion of sorts, Chapter 11 gestures toward renewal and restoration.

The historical moments that I study, the personal experiences that I recount, and the literary texts that I examine are each a "spot of time," a concept that combines event and memory and time and place. From William Wordsworth's

The Prelude, this paradoxical idea offers me a method to connect person, place, and planet in an era of climate change.[1] In nodding to mythology, I remind us that some stories are buried deep in our experience: mythologies recall our origins as a species and speak to us of guiding forces that have long haunted our collective unconscious. To suggest dawning awareness of our individual and cultural entanglement with climate change, after Chapter 2 I gradually add to the historical narrative not only personal experience but also increased attention to literary texts, which invite rereading and offer insights about how water touches everyday life.

Using multiple perspectives to tell a story about water in the Anthracite Region models how we should confront climate change, which troubles us at multiple levels—personal, cultural, economic, and technological—and in scale and scope stretches from the everyday to the cosmic. The only way that I found to capture our predicament was to use as a touchstone the flows of water in a place where U.S. fossil fuel production first found its footing, a ground that sparked industrialization elsewhere and changed water cycles locally and globally. That I am native to this spot raises thorny questions for me about responsibility for then and now.

Working Watersheds challenges disciplinary expectations of history and literary studies, especially chronology and close reading. We now know that the human story is embedded in cycles of earth systems; to write about these cycles is to rethink the centrality of linearity and offer close *and* distant readings of events. To represent nested cycles, *Working Watersheds* tries out different narrative forms and searches at varying depths for ways to see change. To mimic water cycles, chapters tell place-based stories that move back and forth in time and space. As I point out in the above epigraphs, we need to see with both eyes: hot and cold, local and global, form and content, literature and history. After all, Pennsylvania is not only surface, it is what lies beneath.

The changes that I describe are not so much chronological as cyclical, which lends itself to a topical approach. To grasp for a strict linearity within chapters is to miss, ironically, the long view of change that unfolds across the book. From a wide-angle perspective, *Working Watersheds* moves from prehistory, to first mining, to the anthracite industry's troubles and collapse, and then returns to fossil fuel production in discussing the region's natural gas boom. This movement is a design feature of the book: change happens in linear *and* cyclical ways over short *and* long spans of time. Things change moment to moment, but they also cycle round.

The book's first-person stories, historical narrative, and literary analysis draw their meaning from one another; they do not stand alone. As I say in the introduction, the book asks readers to attend to paradox, irony, and juxtaposition.

In *Working Watersheds*, I use methods and techniques of creative writers to think about intersections between nature and culture over time. Echoing Ivan Jablonka in *History Is a Contemporary Literature* (2018), I urge historians to pay attention to form as much as to content as they write.[2] Historians are writers as much as they are researchers, so I offer *Working Watersheds* as a fresh way to widen conversations about how to write history in this new geological epoch, the Anthropocene. If nothing else, this book might trigger debates about the nature of history, the significance of literature, and the importance of linking person, place, and planet.

Bill Conlogue

Timeline

1812–1815	British blockade of Philadelphia during the War of 1812
1829	Delaware and Hudson Canal Company starts regular shipments of anthracite to New York
1829	First successful locomotive run in U.S. at Honesdale
1831	First underground mine in Carbondale
1841	First iron furnaces completed in Slocum Hollow, later Scranton
1850	First underground mines in Scranton
1852	First coal breaker in Scranton at Diamond colliery
1866	City of Scranton formed through combination of Providence, Hyde Park, Scranton
1868	First major mine disaster in Scranton at Diamond mine; lift cage fails; 13 killed
1869	Avondale disaster; 110 men and boys killed
1870	Mine safety laws passed
1872	Silk manufacturing begins in Scranton
1877	Coal strike/violence in downtown Scranton
1878	Lackawanna County formed from Luzerne County
1881	Scranton Steel Company founded
1886	First successful electric trolley system in U.S. begins in Scranton; electric street lights
1890	Scranton Lace founded

1900	Anthracite miners go on strike
1902	Anthracite miners go on strike; action leads to federal intervention
1917	Anthracite mining reaches peak production at 100 million tons
1930	143,000 people reside in Scranton
1959	Knox mine disaster
1962	Old Forge borehole drilled to maintain level of anthracite mine pool
1966	Last underground mine in Scranton closes; the city forms a sewer authority
2006	Cabot Oil & Gas drills its first test well near Dimock, Susquehanna County
2023	76,000 people reside in Scranton

Watering Place

Returning from our tour of the Clark vein, we file behind billboard maps of coal seams and step into the shack at the base of the shaft. As we huddle inside, glancing at walls of yellowed news clippings, our guide hands us souvenir miner's certificates, with 1902 pay rates on the back. Out in the tunnel, arriving tourists emerge from mine cars that will return us to the surface. While someone shepherds the newcomers to the maps, our guide ushers us to the siding, warning us to watch our heads as we duck inside the cars. I slip into the lead carriage. The guide slams doors, settles up front, and speaks into a walkie-talkie. The cars jerk forward, wire rope pulling us along, around a bend, up an incline. The tracks glisten; water spatters the compartment. When a child wonders aloud how long the trip will take, no one answers. Soon a sliver of sky appears, grows, widening to become the mouth of the mine, which draws closer. Finally, we slide into sunshine, a hot, humid August afternoon . . .

Working Watersheds

Introduction

Water

Tell me, O Muse . . .

Keepers of literature and history, the Muses live beside the springs of Hippocrene, on Mount Helicon. Former water spirits, the nine daughters of Zeus and the Titaness Memory bridge present and former worlds. I turn to them, these reminders, to remember water cycles that frame civilization's linear rush to ruin. Our incessant mining damages land, poisons water, and contributes to climate change; reclamation hardly heals the harms. Thinking about water, however, may humble us enough to reimagine energy production, which now entangles us in our own creations. Water slips from our grasp, disappears into thin air, and retreats into blocks of ice. It also writes and erases, sustains and destroys, recalls and renews. Attending to its cycles should steer us to alternative energies, beyond a mining past resurrected in the current stampede to natural gas. Despite the conjunction in the name, hydrocarbons are hydrophobic; they don't mix well with what makes life possible: water.[1]

I search below for the industrial history of water in the Lackawanna Valley of northeastern Pennsylvania. The narrative hides in plain sight in coal mining and textile making, bodies and books, railroading and manufacturing. My guiding assumption: a story about one breakdown might forestall another.

"Working" means imminent collapse and to avoid collapse. Among coal miners, the word "working" names the noise a tunnel makes before it falls apart. Propping up roofs, "dead work," may make the mine safer but never safe. To study working watersheds, I explore where I live and work, the Lackawanna Valley, home to Scranton, once the anthracite capital of the world. I sketch how tainted water circulates here to tell fresh stories about the region's environmental history, which is a cautionary tale for the nation. A haphazard, violent, and slow-moving geoengineering project, coal mining remade the valley, turning much of it inside out.[2]

Reading coal measures mainly as sellable and combustible, nonminers miss what miners know: the earth has a long history of violent change. Using working memory, or recall in action, a miner reads this record carefully because his life depends on it. In confronting the past, such as how to detect and cope with a fault, his working memory draws on knowledge in his short-term memory to assess the danger and decide what to do. Putting what he retrieves to work, he cuts coal, collecting geologic memory, which includes the planet's climate history. What he sends to the surface, though, others burn without thought.[3]

About upstream and downstream, erosion and deposition, a watershed, with its fuzzy edges, marks limits that renew life, potentially endlessly, and for some represents a sustainable social and political community. Complicating bioregional hopes, natural and artificial watersheds shadow one another in the Lackawanna Valley. Its surface river empties into the Susquehanna near Pittston; subsurface streams fountain from a borehole in Old Forge. Threatening aquatic life, the hidden watershed poisons parts of the visible one, and from early days, a coal cartel dominated life in each. Unsettling the dominant narrative of humans as masters of the world, stories of these basins, about people, coal, and water—impure, hybrid tales—follow the Lackawanna River meandering through free and company towns, trace a toxic watershed cupping a recovering one, and think about how a valley warms homes and the world. Cycling through places and people, water in these stories saturates surfaces, subsurfaces, and spaces in between and beyond.[4]

Wrestling anthracite from the earth anchors many stories about the Lackawanna Valley. Historians tell of struggles tied not only to mining practices and mechanization but also to immigration, unionization, and economic expansion and stagnation. Behind these tales stands another story: although men mined a lot of coal, they mined a lot more water. Widening the region's narrative, the water story reimagines the human-coal drama as a water-energy cycle and not a straight path from abundance to exhaustion. In the water story, anthracite mining floats on loops of rain, flood, and drought.[5]

Energy production and water circulation in the Lackawanna Valley created positive feedback loops of technological and environmental change. For over a century, coal companies fought to keep mines dry; operating largely unseen, their ever more complex and costly dewatering devices shaped company choices. Despite their best efforts, however, anthracite engineers increasingly found themselves unable to manage tensions between mining coal and mining water. By the mid-1950s, dewatering had become unsustainable: for every ton of coal raised, collieries pumped to the surface forty tons of water. Similar feedbacks haunt us today as companies shatter shales for gas, claw sands for oil, and level mountains for coal. In a world threatened by climate change, these loops in the water-energy nexus threaten to strangle us all.[6]

The industrialized world is responding to climate change the way that anthracite coal companies responded to water. Colliery owners groped for technical fixes—work-arounds, new equipment, more efficient practices—that ultimately couldn't reverse the vicious cycles inherent in water-coal connections. Collapse came slowly and painfully, filled with adaptations and hopes, successes and disasters. Redirecting our response to climate change, which threatens culture change, requires a change of mind; to rethink what we do, we need to hear the stories that water tells. To interpret its tales, I turn to history and literature as guides.

To understand the water-energy nexus, I look to an early contributor to anthropogenic climate change—anthracite coal mining—for three reasons: anthracite jump-started the industrialization of the United States; as a native of northeastern Pennsylvania, I live in the aftermath of mining; and fossil fuel extraction has recently returned here in the form of natural gas drilling. Meshing these narratives to offer new perspectives on the Anthracite Region, I rely on paradox, irony, and juxtaposition.[7]

My off-farm employment began with a stint at Pleasant Mount Welding, a manufacturer of wastewater treatment systems. Not long later, I landed a job in Simpson making heat exchangers at Doyle & Roth, a Manhattan firm formerly headquartered in the Standard Oil Building, which also once housed the Modern Language Association. Today I teach writing and American literature at a university that pumps mine water to cool the building that shelters its architecture program.

We often rework our past workings. In the last decade, the natural gas industry has stepped into the afterlife of coal mining. Located beside the Lackawanna River and adjacent to downtown Scranton, Steamtown National Historic Site collects steam locomotives that once both burned and hauled anthracite; railcars filled with frac sand now regularly roll north through the park. Built for the coal industry, the former railyard in Carbondale has found new life as a transfer station for natural gas companies. Trucked from there to Susquehanna County, frac sand enables drilling crews to recover gas from shale. Railbeds only recently converted to hiking and biking trails now double as routes of natural gas pipelines.[8]

Tracing local flows of water helps me know the history of the Lackawanna watershed, which reveals a wider world awash with ambiguity. The uncertainty in choosing between more than one meaning, ambiguity names a gap that bridges possibilities, a fact represented in the hyphen of the water-energy nexus. Understanding how the nexus contributes to this uncertainty helps me—and maybe others—to see home more clearly. I assume that I make decisions on firm footing, but more often than not, I must choose while caught in cross-currents of doubt. Most issues are not, after all, black and white. About fluidity, porosity, and transformation, *Working Watersheds* seeks answers to the question that has shaped my teaching and writing: how have we been at work in the world?

Searching for answers, I move with caution, remembering Narcissus, who fell for his own reflection. Had he looked beyond his face, he might have saved himself. Like Narcissus, I sometimes fail to find the depths below me; I know then only surfaces. No matter whether I see my expression or a deep, watery world, I understand that a single touch can distort and disrupt both the surface and what lies beyond, sending ripples far and wide.

Taped to my office door are the words "Pay attention," which sum up what I hope to teach students. Novelist and short story writer David Foster Wallace explores this idea in his 2005 commencement address at Kenyon College. He opens his talk with a story about three fish. An older fish says to the younger two, "Morning, boys, how's the water?" As the youngsters swim on, one wonders to the other, "What the hell is water?" Urging graduates to pay attention, to challenge routine, to practice empathy, Wallace concludes that the "real value of a real education" is "awareness of what is so real and essential, so hidden in plain sight all around us, that we have to keep reminding ourselves over and over: 'This is water. This is water.'" Writing and revising, I tell myself, "This is water. This is water."

Everyday things too often escape my notice because they are so obvious, so much part of my world. A driver, I pay at the pump, get the oil changed, and visit carwashes, often with little thought beyond my bank account. But the more I learn about climate change, the more I think: small acts have big consequences. Soon I may know what so many others already experience, displacement without leaving home. I imagine that miners sensed this 100 years ago as they endured mine subsidence, mine fires, and near-daily dramas of disability and death, plus flood, drought, and disease.[9]

Until recently, it had never occurred to me to wonder much about the infrastructure of home: the pipes, poles, and power lines that make my mundane acts so easy. A record of land transformation, infrastructure also hides in the hyphen of the water-energy nexus. Although it separates and conjoins, the line fails to acknowledge that the nexus names a cyclical flow or that infrastructure both speeds and slows its motion. In northeastern Pennsylvania, water determined the infrastructure of the industry that shaped, wounded, and ended many lives. The first large-scale alterations of land here established a transportation monopoly: a single mine-to-Manhattan rail-canal-river loop. All around me I see the constructions that followed: streets, tracks, bridges, and highways. What do they tell me about where I am? I push up the thermostat, but what do I know about the system that warms my house? I put on a sweater, but what do I know of its making? Climate change demands that we ask good questions, because if we don't, we're dead. Below, I ask questions that I invite you to answer, if only tentatively, about your own waterscapes.[10]

Mining usually means a boom-and-bust economy, industry-friendly oversight, and polluted land, air, and water. As anxious workers suffer a damaged place, others elsewhere worry less and profit more. Returning as natural gas drilling, mining these days threatens not only another replumbing of the place, with gas wells, pipelines, and power plants, but also a replumbing of the planet in our tinkering—knowingly—with the global water cycle. Burning fossil fuels defiles waters everywhere; living in the Anthropocene, I see the world in the Anthracite Region, a land marked, unremarked, and remarkable, a place diminished but not done.[11]

In many economies, some places get mauled so that others might go on, and abandoned mines and gas wells have long raised questions about right and wrong. The questions, though, take on more urgency today because industrialized nations demand more and more sacrifice just to run in place. Although recovering, the Lackawanna Valley lives with legacies of coal mining. In 2018, threats of mine subsidence forced Lackawanna College to fill voids beneath classrooms in downtown Scranton. In the same year, the Pennsylvania Department of Environmental Protection worked to put out the Powderly Creek coal fire in Carbondale, which officials had been trying to extinguish since the 1980s. Meanwhile, adjacent Susquehanna County accumulated legacies of natural gas drilling. In response, energy companies shrugged: sorry for your inconvenience, but you must sacrifice yourself for the greater good, which, of course, is carefully calibrated to our bottom lines.[12]

When mining dominated life in the Lackawanna Valley, water enabled and confounded coal companies in their efforts to extract profits from the earth. To survive, operators warily watched the quantity and quality of water that circulated through the valley. To not only wash and market coal but also to power pumps, colliery owners redid the natural water cycle. As they dug deeper, water forced them to invest more time, attention, and money to fight floods, survive droughts, and satisfy consumer aesthetic demands. Wrestling with water eventually undid the industry, a reminder that when water and fossil fuels clash, water wins, always.

To contend with water, coal companies grafted artificial hydraulic systems onto natural ones, creating new circuits in the local water cycle. The first circuit percolated "naturally" as acid mine drainage; the second flowed through reservoirs and channeled parts of rivers and streams; the third, more obviously human made, moved mainly in pipes throughout the region and within homes and businesses. Evolving as a single system, these natural and artificial circuits intertwined with global water cycles, themselves often enough hybrids. Not only do people here live with a ghost anthracite infrastructure, degraded land, and a partially dead river, but long-burned anthracite, now atmospheric carbon, continues to affect how water flows around the world, increasing the intensity of floods and droughts here and elsewhere.

At the height of the anthracite industry (1890–1920), the valley confronted a perfect storm of conflicting water and energy demands. An increasing population, supported largely by jobs in mining and textile making, demanded clean drinking water, efficient waste disposal, and coal for home heating; at the same time, to keep operations humming, industries, railroads, and power companies required ever more water, ever more coal. Meanwhile, collieries poured an ever-greater amount of mine water into streams, typhoid killed people, and more and more sewage polluted the Lackawanna River, whose waters reach the Atlantic by way of the Susquehanna River and Chesapeake Bay.

I write on this side of geologic upheavals that have made and remade this valley. I write too about memento mori, remnants that recall a story that water still writes, a memory that whispers possibilities and cautions. Putting these discoveries into words recalls—partially, minimally—what was, which is often a site of catastrophe, whether sudden or glacial. Just as fossils, both material and linguistic, call attention to spots of time, or place-moments to think with, so do a buried valley, an abandoned building, and fading print.[13]

Historians confront a problem: climate change enlarges the scale of the narrative. They can no longer tell only a human drama, which, we now know too well, is caught up in a planetary, if not universal, one. Faced with such large swaths of time and space, writers must reach for new methods to tell coherent tales that trace meaningful interrelationships among person, place, and planet, organic and inorganic, and animals, plants, and soils. Testing the relevance of local history to this expanding story, I follow not only the Lackawanna River but also water in its watershed, above and below ground and beyond. I write about home to say something about the world, our home.

But writing this is a problem: how I know what I see keeps changing, often leaving me perplexed, which is why I seek to know working watersheds through multiple prisms. From the deep-time perspective of water, this postindustrial valley is multiple places of perpetual change: rains erode soils, streams cut new channels, and water vapor spurs plant growth. As I write, I discover this land as multilayered, always unfolding, of varying depth, and threaded with human and ecological contexts, some immediate, others distant, some significant, others less so, depending on the questions I ask. To sharpen my view, I look aslant. Acknowledging the grayness of experience, *Working Watersheds* invites connection, interpretation, and conversation, inching us closer to the truth, or so I hope.[14]

Poets remind me that a search for the truth can end in ambiguity, even when one looks aslant. In 1917, the peak year of anthracite production, Robert Frost wrote "For Once, Then, Something," which begins with the speaker noting how "Others taunt me with having knelt at well-curbs" (1). Although, like Narcissus, he usually sees in wells only his own reflection, "a shining surface picture / Me myself in the summer heaven, godlike" (4–5), one time he discovers "beyond the picture, / Through the picture, a something white, uncertain, / Something more of the depths" (8–10).[15]

The vision does not last because "Water came to rebuke the too clear water" (11): a single drop ripples the image, blurring and blotting out whatever it was. The speaker wonders what he had glimpsed, "Truth? A pebble of quartz?" (15). It could have been, "For once, then, something" (15), or it could have been nothing but a stone, some thing. He perceived either more of an abstraction, the world of ideas—perhaps Plato's reality—or more of a concrete, material world. Despite looking for the bottom from a new angle—"with chin against a well-curb" (7) and right to the light (2)—he draws from the well only fleeting, ambiguous understanding. Whether he has seen into the nature of things or found an idea about it, no matter what, water offers access to it, shapes it, and erases it.[16]

A fluid history, one that traces transformations, this book meanders in its flow. I use "meander" and "flow" because their senses overlap. As a noun, "meander" means a winding path, especially a labyrinth; as a verb, it is to follow an intricate course and to wander casually with no pressing destination. The word's origin, the Meander, a river in present-day Turkey, follows a crooked path to the sea, long after it abandoned its ancient port city.[17]

"Flow" comes from the Old English *flówan*, which is akin to the Old High German *flouwen*, to rinse or wash. The word has roots in the Latin *pluere*, to rain, and the Greek *plein*, to sail. The English word means "to issue or move in a stream . . . circulate"; "to derive from a source." In maybe the most important definition for this study, however, "flow" means "to deform under stress without cracking" or fracturing, used mainly of "minerals and rocks," but is just as helpful in thinking about people and places.

About survival and adaptation, "flow" sums up the aftermath of mining in the Lackawanna Valley, which has been disfigured but not cracked, at least not completely. The pressures of climate and energy change will likely deform it further; how individuals and communities cope will determine whether they shatter. Given the grip of global warming, maintaining local flow may be the most positive possibility. If changing form is survival, then defending human integrity and dignity in the Anthropocene will be about learning how to build lives, communities, and infrastructures that won't crack or split under stress.

The Water-Energy Nexus

The nearly limitless light of the present powered ancient civilizations; modern ones burn the limited light captured in ages long past. In early agrarian worlds, human work did relatively little widespread ecological damage, but in mining the underworld, our largely urban work today degrades lands and waters everywhere. Although the sun powered work life into the nineteenth century, since then, fossils have energized industrial systems so much so that the earth's waters should now hold our attention.

Confronting global warming requires rethinking the dominant water-energy nexus. With no origin or end, the hydrological cycle loops on and on, driving untold volumes of water, transforming liquid to vapor and back again. People have always done what they could to capture, redirect, and use water, but only since the advent of coal-fired factories have humans seriously affected the cycle itself. Like everyone else, I'm caught up in this circuit of waters moving in time immemorial, rivers flowing across geologic time, and mining happening within human time. Unable to stand outside these, I can't know the water cycle completely, but I learn what I can.[18]

I first saw the phrase "water-energy nexus" defined in a 2014 publication of the U.S. Department of Energy, *The Water-Energy Nexus: Challenges and Opportunities*. The report points to 2012 as a watershed year: drought curtailed production at power plants, Hurricane Sandy ravaged water infrastructure, and fracking further complicated discussions of water and energy. Fretting about energy security, a form of national security, the writers mainly urge more efficiency at its intersection with water. But even as others now rethink the nexus, some focus on only one side of the hyphen, hardly mentioning the other. Not to critique today's relationship between water and energy is to refuse to imagine whether the link may be more fragile than many assume.[19]

Water alarms policymakers worldwide. Global use of fresh water jumped in the twentieth century at "several times the rate of population growth." In 2008, the *Wall Street Journal* declared water the twenty-first-century equivalent to oil. In 2014, the World Economic Forum named water one of the top-three threats to the resilience of the global economy, sparking urgent discussions of the water-energy-food nexus. In 2018, world leaders worried enough about water that the United Nations named 2018–2028 the Water Action Decade.[20]

Intensifying rains and prolonging droughts, climate change pressures a choke point in the nexus: only so much water is available at one time in places of power production, just as there is only so much energy available at one time in places where people and industries pump water. Over 40 percent of U.S. fresh water withdrawals go to thermoelectric power production. Mainly due to population increases and tighter regulations, between 1996 and 2013, U.S. water and wastewater treatment plants demanded 30 percent more power. To keep coal mines dry, companies pump water from them, but to mine natural gas, they pump water from rivers, force it *into* the earth, and collect what flows back, only to recycle this wastewater or force it back underground elsewhere. Building more infrastructure to move more water and energy to where it's needed may magnify the climate problem because it will take a lot of water and energy—and money—to build and operate it.[21]

Human activity disrupts water cycles near and far. Due to rising salinity in the Nile delta, Egypt faces the prospect of uninhabitability by 2100. Siphoned for irrigation, the Aral Sea is now one-tenth the size it was in 1960. In the U.S., a desert threatens the Salton Sea—once a jet-set resort, now a poisoned sump—and multiple demands have long kept the Colorado River from reaching the ocean. In Florida, algae from farm runoff and lawn fertilizers threaten to kill Indian River Lagoon, the "most biologically diverse waterway in America," and toxic blooms choke bodies of water across the United States, including reservoirs.[22]

Stresses in the water-energy nexus take many forms. In 2012, frackers outbid farmers at a water auction in the Colorado River basin, which has long walked a tightrope between rural and urban needs. During the California drought in 2011–2019, the state chose energy over water: when the governor ordered that urban water demand be cut by 25 percent, he "left untouched" the oil industry. In 2013, Florida sued Georgia for taking too much water from the Chattahoochee and Flint Rivers for farmers and metro Atlanta. Seeking to limit Georgia's draw, Florida blamed lack of water for the 2012 failure of the oyster industry in Apalachicola. In 2016, Roseburg Forest Products notified a lumber town at the base of Mount Shasta that it needed to find another water source. For a half century, the city had paid the company one dollar each year to use Beaughan spring, whose water the company had decided to sell in bottles, presumably made of plastic, a petroleum product.[23]

The Weirdness of Water

A hero returning home from the Trojan War, Menelaus, husband of Helen, bewails his ships becalmed beyond Egypt. To find fair winds for the journey to Sparta, he discovers that he must seek Proteus, a shape-shifter, an attendant if not son of Poseidon. Unlike Teiresias or Pythia, Proteus, a reluctant guide, must be captured and held before he will speak. As Menelaus and his men grip the old man, refusing to let go, he swiftly changes form, from lion and panther to water and tree. Fixed, finally, he tells the truth: to reach his kingdom, Menelaus must first return to Egypt to offer sacrifices to the gods. Nodding to the Deathless Ones, Proteus teaches Menelaus that to steer a course for home, he must humble himself before perennial powers.[24]

A shape-shifter, water possesses properties found in no other liquid. About condensation, circulation, and evaporation, it seeks its own level and combines readily with itself and other substances, though not with oil. And it as readily separates from them, making it a "universal solvent," one that can distribute nourishment and oxygen throughout the living. Depending on how it goes, this can lead to life, illness, or death. Its weirdness, however, allows it to store and transfer energy, making it a key element in electricity generation. Even as it helps and hinders, enlivens and kills, and comes and goes, water varies in duration and amount from here to there, from time to time. It respects no geographic, social, or political boundaries, and though humans may capture it behind dams, some slips through the cracks or gets carried off in evaporation. Water cannot be destroyed, only transformed into different states, solid, liquid, and gas. Although we cannot destroy water, we can deny access to it, make too many demands on it, and muck it up enough to make it unusable, if not harmful. In doing so, the life giving becomes life destroying. Both nature and culture, water helps me rethink opposites; it's less either/or than both/and.[25]

The weirdness of water matters because burning fossil fuels disrupts world weather, which is wind and water. Human-caused climate change—along with extinctions and pollution, not to mention world-ending weapons—offers evidence of our effects on the world, effects so great that the earth's geology records our presence and passing. Some scientists have named this new era of change the Anthropocene and date its start to improvements to the steam engine in 1784, which is about when life in lakes began to change. Altering since then the "geochemical cycle in large freshwater systems," humans now use "more than half of all accessible fresh water." Other signs of the Anthropocene: radioactive fallout from the 1945 Trinity test, which marked the geologic record worldwide, and the "Great Acceleration" of human activity that has happened since World War II. Closer to home: when men hacked out coal, they carved the human presence within the geologic record, which responded by bleeding acid.[26]

Bad news keeps piling up. In 2018, the International Panel on Climate Change reported that within the next few decades, global warming will be widely and painfully felt. A month later, the Trump administration sounded an alarm with the *National Climate Assessment*, which claimed that climate change was no distant event: it was now. In 2019, the United Nations proclaimed that human land use worsens climate change, which further degrades land, pressuring food production. In 2021, the International Panel on Climate Change warned that "continued global warming is projected to further intensify the global water cycle, including its variability, global monsoon precipitation and the severity of wet and dry events." Mitigation means remaking the water-energy nexus now.[27]

Climate Change

Water and carbon dioxide drive climate change. A greenhouse gas, water vapor blankets the earth, keeping it from becoming a cold, barren rock. Carbon dioxide regulates this effect, making CO_2 the key to climate change. Adding CO_2 to the atmosphere, which thickens the earth's blanket, threatens to suffocate all life. The more that people burn fossil fuels, the more carbon heats the atmosphere; the more that carbon heats the atmosphere, the more energy we need to cope with the consequences: severe drought, rising seas, and heavier rainfalls. Altering the earth's hydrologic cycle, this positive feedback loop makes climate change a "wicked problem."[28]

Climate change now accelerates across long and short timescales. Scientists predict that by 2100, the average worldwide surface temperature will shoot up between two and five degrees Celsius, "a faster rise than any experienced during the last million years." In 2018, the United Nations reported that global temperatures are rising faster than first thought; between 2030 and 2052, temperature averages may be more than one and a half degrees Celsius beyond preindustrial levels. A few months later, scientists announced that Alaska was experiencing "never-before-seen melting and odd winter problems, including permafrost that never refroze." The Arctic has lost almost all of its old ice, researchers discovered, and has experienced "increased toxic algal blooms, which are normally a warm water phenomenon." In 2019, scientists again warned that things are happening faster: a rare summer heat wave melted Greenland glaciers at a clip seen only once before, in 2012. These announcements added to existing evidence that "ice-core records suggest that half the warming in Greenland since the last ice age was achieved in only a few decades."[29]

Even as people grew more aware of climate change in the 1990s, industrial economies poured gas on the global warming fire. During the last 30–40 years, carbon emissions have exploded. The world now annually burns 400 years' worth of accumulated plant and animal life. Half of all CO_2 belched from burning fossil fuels between 1751 and 2010 spewed into the atmosphere after 1986. Registered another way, one-quarter of CO_2 emissions between 1874 and 2014 happened in 15 years, from 1999 to 2014. By 2015, the level of CO_2 in the atmosphere had ballooned beyond preindustrial levels by more than 40 percent, and it could go another 40 percent higher by 2100. This is no sane way to confront a climate crisis.[30]

Climate change has increased the odds of extreme events. In 2018, heat waves rolled across the earth, forest fires flared in Europe, and heavy rains flooded East Coast towns. By mid-August, forest fires in the United States had burned almost 9,000 square miles, "about 28 percent more than the 10-year average." Warmer-than-usual river water shut down nuclear power plants in Europe, and scientists warned of "a future of cascading system failures threatening basic necessities like food supply and electricity." The 2017–2018 winter season was the warmest on record in the Arctic, where "weather stations averaged 8.8 degrees warmer than normal." The summer of 2023 was the earth's hottest ever recorded.[31]

If climate change is water change, people—especially in the industrialized world—will use ever more energy to survive a global water cycle gone haywire. Much of this energy may go to pumping water during floods and droughts, building water-tolerant infrastructure, and cooling power plants, businesses, and homes. The more these activities depend on fossil fuels, however, the more the earth will warm, further disordering the water cycle, which will force people to use more energy. These feedbacks will likely enhance the power of fossil fuel companies, making them less likely to combat climate change, which can only lead to an acceleration of climate weirdness. At the heart of the problem lurks a paradox: industrialization has not only insulated many people from extreme weather events but also increased their vulnerability to weather catastrophes.[32]

Although tensions between water and fossil energy drive climate change, water will win, a fact foreshadowed locally. The water-coal story here begins and ends with a river threatening to stop men from mining. In 1822, workers in Carbondale jerry-rigged a waterwheel to keep an icy Lackawanna from flooding their coal quarry. In 1959, an icy Susquehanna, near its confluence with the Lackawanna, broke into the Knox workings, all but ending deep mining in the northern anthracite field.[33]

From the vantage point of the Anthropocene, the heat wave that roasted the Northeast in 1917 appears prophetic. In this peak year of anthracite production, high temperatures killed at least 8 locals. Nearly 200 perished in New York and 143 in Philadelphia, and in Boston, 13 died in 24 hours. Just as prophetically, as a massive thunderstorm cooled New York, lightning strikes killed 4.[34]

Recalling the history of the water-energy nexus in the Lackawanna Valley reveals the degree to which coping with water contributed to the collapse of anthracite mining. This local history foreshadows how a changing global nexus might affect all life. The intersection of deep coal and too much water undid the anthracite industry, which cratered the regional economy, ending other businesses, forcing an exodus of people, and leaving a plague of lingering environmental problems. The coming coalescing of peak water and peak oil and gas will undo fossil fuel economies, a fact that climate change hastens. This time, though, there's nowhere to run.[35]

With water saturating works of mining and textile making, the history of water and coal in the Lackawanna Valley teaches me much about how increased water abundance and scarcity might play out as global temperatures rise. Sometimes energizing, sometimes dampening lives, water presses upon my awareness because I see that climate change forces the world to confront problems that bedeviled anthracite workers: high water, bad water, and no water. At the end of my book *Undermined in Coal Country*, I hoped for a greener world. As I write now, I remember that any green world is always also, and first, a blue one.

1

Writing Water

Recalling the removal of minerals and money from one place to other places, the extraction narrative often focuses on farsighted entrepreneurs, venture capitalists, and ambitious workers. Explaining the rise and fall of the anthracite industry, the local version also explores immigration, labor strife, and boom-and-bust economies. This story, which privileges complex human interactions over time, makes the present meaningful: if nothing else, events between 1829 and 1959 still define today's Anthracite Region, a distinct geography whose identity derives from an industry that has all but ceased to exist. People made the area a place by hollowing it out, separating its parts, and shipping pieces elsewhere to be reduced to ash.

This past often persists as myth. During the 2016 presidential campaign, candidates who visited Scranton portrayed northeastern Pennsylvania as a coal-mining region, despite residents knowing that the last commercial deep mine in Lackawanna County had closed 50 years earlier. When Donald Trump vowed to bring back coal mining, many scratched their heads in wonder. Going with the story behind this promise, a city businessman hosted a rally billed as "Miners for Trump"; on the same day, others rallied under the same name in Philadelphia and Pittsburgh. When Trump next visited the region, to stump for Lou Barletta in his Senate bid in 2018, the president promoted another myth, "our beautiful Pennsylvania clean coal," again reminding his audience of the region's extraction narrative. When Joe Biden ran for president in 2020, he played to this history when he included in stump speeches that his great-grandfather had worked as a mining engineer in Scranton.[1]

Remembering that violence haunts extraction, the subsidence narrative tells the aftermath story, often of deaths, depopulation, and deepening distress. Anthracite mining killed nearly 40,000 workers outright, wounded countless others, and condemned thousands to black lung. The ruin of the industry left many unemployed, forcing most to flee the region to find work. Those who stayed suffered low wages, long commutes, and chronic underemployment, a litany that continues: from 1992 until 2022, Scranton lingered in official financial trouble.

This description of extraction-subsidence narratives tells only the human half of the story. What of the other half, the nonhuman? Overturning seams of time, scarring the surface, and leaving behind piles of waste rock and scores of polluted streams and rivers, extraction reshaped this land. Mining killed nearly all life in the Lackawanna River, drove most wild animals from the watershed, and long robbed those that remained of much habitat, especially tree cover. Carbon dioxide, a spectral effect of a largely undone industry, still contributes to transforming the earth. To know the full extraction-subsidence story, which culminates in climate change, one must tread water. Life taking and life sustaining, water was for coal companies a nemesis and a necessity; for communities that lived with mining, it was a liability and life. Water ghostwrote the extraction-subsidence narratives that dominate histories of the Lackawanna Valley.

Extraction-subsidence narratives unfold within a larger pattern, the water-energy nexus, a global cycle that interlaces water and human work. The most common extraction-subsidence stories move in one direction, chronologically, from origins to ends, whether tragic or triumphant. Read against the water-energy nexus, however, extraction-subsidence narratives manifest as cyclical, loops that circulate locally, extend globally, and tend to go unnoticed, until catastrophe strikes. The loopy story of how coal mining altered water circulation in the region is one tale in the weird history of how burning fossil fuels has been changing the earth's hydrologic cycle. In our age of oil, nearby natural gas drilling is rewriting this story again.

Water wrote first drafts of local extraction and subsidence narratives. A deep record of deformation, the geology of the Lackawanna Valley reads like a series of slow disasters. In swamps of the Pennsylvanian series of the Carboniferous period, about 300 million years ago, trees and grasses captured light, died, and collected under water. The long piling pressed out the moisture, leaving carbon, which hardened. Continents then collided, folding the land, rolling beds of coal, forging Pennsylvania's ridge and valley topography, which water eroded. After the land masses separated, a sea filled the folds; its subsequent disappearance left a salt lake, whose bed collapsed, creating the valley. Advancing and retreating ice later buried its lower end in glacial till, over which now flows the Susquehanna River. Burning anthracite, or stone coal, the hardest coal, almost pure carbon, released heat and light that fired, briefly, furnaces that not only made goods and warmed homes but also triggered the feedback loops of climate change that have, ironically, made ancient geologic catastrophes newly present.[2]

Just as people do, water composes palimpsests, new writings that replace older ones. In the rewriting, past scribblings don't simply disappear, they often bleed through the new. Gouging gullies, eroding hills, and saturating soils, water repeatedly marks land; writing over this story, mine workers build breakers, sink shafts, and extract coal. Reading lands and coal measures, geographers, geologists, and historians discover meaning in stories of violence, indifference, and change. Even as this happens, flood, rain, and snow rewrite it all. Long intensifying, these loops of inscription now warm the earth, threatening it with loss of meaning.

Such loss has happened. First used in 1825, the word "palimpsest" comes from the Greek for *palimpsestos*, or "scraped again." An ancient palimpsest: about 750 million years ago, mile-high ice pulverized layers of rock two to three miles thick, erasing a billion years of the earth's history. The sediment washed into seas, settled at the bottom, and disappeared into the mantle, only to resurface as volcanic magma. The scraping left the Great Unconformity, a geologic gap that links layers of largely lifeless time below with horizons of living beings above. About 450 million years later, anthracite formed from strata of fossilized lives.[3]

The region's ecological past has been unintentionally edited from much anthracite history. Writers and revisers, historians study the Anthracite Region from a variety of human angles, mainly economic, social, and industrial. While Burton Folsom focuses on male entrepreneurs in *Urban Capitalists* (1981), John Bodnar interviews everyday folks in *Anthracite People* (1983). In *The Kingdom of Coal* (1985), Donald Miller and Richard Sharpless offer a comprehensive history of the anthracite industry, and Harold Aurand explores the region's unique industrial and social past in *Coalcracker Culture* (2003). In *The Face of Decline* (2005), Thomas Dublin and Walter Licht trace how communities and individuals coped with the waning of the coal industry, but Richard Healey recounts its rise in *The Pennsylvania Anthracite Coal Industry* (2007). Just as Karol Weaver uncovers a history of formal and informal medical care in *Medical Caregiving and Identity in Pennsylvania's Anthracite Region* (2011), so Robert Wolensky and William Hastie unveil connections between organized crime and coal mining in *Anthracite Labor Wars* (2013). Using a multidisciplinary approach in *Remembering Lattimer* (2018), Paul Shackel examines the history and memory of the Lattimer Massacre, and through oral histories, Robert Wolensky recalls the twentieth-century unionization of the garment industry in *Sewn in Coal Country* (2020).[4]

The Anthracite Region also has a rich literary history, which similarly focuses on human experience. In 1927, folklorist George Korson publishes *Songs and Ballads of the Anthracite Miner*. Novelist John O'Hara writes about Pottsville in his fiction, beginning with *Appointment in Samarra* (1934). The collection *Coalseam* (2005) includes poems about the region by W. S. Merwin, Jay Parini, and Gerald Stern, and in the anthology *Anthracite!* (2006), Philip Mosley brings together six coal-region plays. Lesser-known short story writer Emily Johnson and poet Paul Kelley add their voices to this tradition. To recall the ecological in the everyday, I place Johnson and Kelley in conversation with anthracite history and bring this history into dialogue with poets Wendell Berry and Robert Frost and novelists Paolo Bacigalupi and Jennifer Haigh.[5]

Recovering the region's ecological past requires this double vision. Writing about water in the context of climate change, I also need to think about geologic history, with its long timescales, and human history, with its relatively short ones. Making things more complex, this larger doubleness is spatial and temporal because as much as human history is about events in time, geological history is also about time in matter. To convey this double nature, I move among history and literature and myth and my experience. In *Working Watersheds*, form is content.

Archbald Pothole

At the onset of the present interglacial period, about 20,000 years ago, the Wisconsin ice sheet retreated from the region, grudgingly, leaving behind gashes, ponds, and boulders. As the ice melted, water pooled on its surface, overflowed, and cascaded into crevices, drilling holes in the land, only to fill them with gravel. About 10 miles north of present-day Scranton, a huge waterfall bored into the coal measures a shaft 38 feet deep and 24 feet wide. A geologic timeline, the gap opened the Holocene to the late Carboniferous.[6]

In 1884, men in the Ridge mines revised this story, extending it into human time. After chipping and shoveling through 600 feet of coal, Patrick Mahon and his workmates set off a black-powder blast that triggered a rush of water and stone. Swept back, the crew scrambled for the surface. When they returned, they found a wormhole to the last ice age. For three weeks, they carted off rock, and then gravel and soil, before seeing sunlight. A "freak of the glacial period," the opening funneled fresh air across the ages.[7]

A memory of what's missing, the pothole, a void, offers access to stacked strata of time that stretch from the Carboniferous to the Anthropocene, 300 million and more years of geologic history. The space unearths stories that not only expose the power of water but also reorient human relationships with it. Making memory manifest, the gap witnesses that ice about 2,000 feet thick once covered the region, reminding observers that other weathers have come and gone, disclosing that the earth's current climate is neither forever nor for long.[8]

People sought talismans to know the hole, the space between now and then. To touch the gap, the curious collected stones that others had mined to reveal it. Tokens of other times, the chips offered evidence of an emptiness that connects eras. The Smithsonian preserved photographic images of the opening, and geologists passed on to students what they knew of it, this "remarkable work of nature." A door to the underworld, the site became such an object of pilgrimage that the landowner spent $500 to protect "that wonderful phenomena against the action of the weather, and also to make it more attractive." His main "improvement": a square of stone high enough to "keep cattle and visitors from falling in."[9]

At first a freak of nature, a curiosity, the pothole soon became just a hole, space to fill. By 1926, few visited, the wall had fallen apart, and people perceived the opening's smooth sides as "more weird than picturesque." Locals choked it with garbage, tossing into gaps between geologic times scraps of human time: "tin cans, old wash boilers, kitchen waste."[10]

In 1940, some saw the hole anew. Faced with a cratering economy—anthracite mining suffered steep declines in the 1920s and 1930s—county officials encouraged people to visit the void. Targeting motorists and campers, boosters reclaimed the gap for public memory by designing a park to preserve an emptiness tied to mining, which, after long hollowing the valley, was collapsing as a viable industry. Two deeds conveyed the site: one for the hole, the other for the "surrounding land." By the 1960s, Pennsylvania advertised the park statewide as a way for travelers to journey back in time. With the interstate highway system making it more convenient for out-of-towners to access the place, planners encouraged them to burn oil and gas to glimpse a gouge in coal. In 1967, state and local officials met to develop the pothole into a park with "hiking trails, a 'demonstration forest' . . . [and] a trailer camp or tent city for tourists." Repairs to an "unsafe" viewing platform, however, stalled because state officials couldn't agree about "underground conditions" beneath the pothole. Today the wall is gone, fading signs tell the story, and woods shade the wound.[11]

Buried Valley

Water wrote other stories. The Wisconsin ice sheet buried the Wyoming Valley under sand and gravel. Cupping the current basin, the buried one channeled an antecedent Susquehanna River. Twenty thousand years after the first vale disappeared, miners hacked out tunnels beneath its ghostly traces. Water filled today, the catacombs constitute a doppelganger watershed. As the Susquehanna flows across the surface, streams of mine water meander below; in between stretches evidence of the ancient river.[12]

Water won't let the buried stay still. Sodden, the old valley breaks surfaces, collapsing worlds above and below, rewriting human history. Between 1872 and 1947, rock strata under the hidden valley failed at least 19 times, suspending mining, threatening scores of men, and killing more than two dozen. A forgotten past rediscovered, violently and repeatedly, the sunken channel upends assumptions beneath stable narratives of time and place. It teaches that no place is one place, or one time, or solid.[13]

In 1897, just over a dozen years after water pressed itself upon the mind of Patrick Mahon, a geologist and mining engineer pointed to the Archbald pothole in explaining to coal operators the existence of the buried valley. He argued that glacial holes cutting through the lost valley likely triggered two mine accidents. In the first, a few months prior, a pothole poured quicksand into mines under the Wyoming post office; luckily no one was at work below, but the sand caused a subsidence that swallowed the building. A far worse accident, however, happened the year after Mahon's fright.[14]

In December 1885, miners chipping at the Nanticoke measures struck a pothole. A torrent of water, sand, and gravel swept men, coal cars, and timbers through the tunnels, trapping workers who fled to higher gangways. In less than an hour, the flow filled the vein, floor to roof, for 2,000 feet. Attempts to reach survivors ended when a rush of water, sand, and culm chased rescuers to the surface. Early reports claimed that 23 men died: drowned, asphyxiated, or starved, their bodies were never recovered.[15]

While anthracite engineers puzzled over glacial potholes, water found other ways into mines beneath the buried valley. Seepage, of course, was constant, and in some places, coal operations disturbed the water table enough to steer water into workings. Mine subsidence drained thousands of small streams, and stripping pits fed runoff directly to underground workspaces. Funneling water to the measures in 1939, a borehole drilled through a subsurface stream caused a gradual settlement that transformed West Wyoming farmland into a swamp. Thirty years and a day after the Nanticoke flood, a cave-in at Miners Mills poured Mill Creek into tunnels. To stop the flooding, crews had to alter the course of the brook. In each case, water spilled from hydrologic time into geologic and human time.[16]

Lackawanna River

Both channel and water, static and in motion, the Lackawanna River slips across measures of time. As flowing water, it represents a movement of time; as a slash in the land, it is a place. When people build walls against it, the river runs as a hybrid of wild and made, nature and culture. A shape-shifter, the Lackawanna mainly makes and marks its own way, recording and revising its passage by rolling stones, eroding banks, and deepening and filling its bed. Although along its edges I may hear it murmuring, the river goes without saying—until flood or drought calls it to my attention.

An Algonquin word, *lackawanna* means "stream that forks." About 62 miles long, the river drains 350 square miles, flows through 23 municipalities, and passes across land populated by nearly 250,000 people. In Ararat, on the Appalachian Plateau, its headwaters slip from "glacial ponds and wetlands." Lake Lorain, Bone Pond, Independent Lake, and Dunns Pond feed the east branch; the west branch carries water from Fiddle Lake, Lake Lowe, and Lewis Lake. After entering the valley at Stillwater, the combined stems cut through layers of Devonian sea floor, Carboniferous swamp, and Permian sedimentary rock. Dropping on average 26 feet per mile, the river then runs 40 miles to the Susquehanna. The confluence, atop the buried valley, marks the cultural divide of a canoe-shaped geography—the Lackawanna and Wyoming Valleys—which itself divides distinct geologies, the Alleghany and Pocono Plateaus. In addition to streams that descend surrounding mountains, more than a dozen mine tunnels add to the river's volume.[17]

A "working-class river," the Lackawanna is emblematic of postindustrial lands everywhere. The river inspired no arts movement, no port city evolved at its mouth, no major battle straddled its banks, and it has no history of inland shipping or international trade. It's not even impressive in length, width, or volume. In almost no time, though, the Lackawanna moved from quiet stream to working water. From the start, coal companies burrowed under it, shoved it aside, and, along with valley towns and cities, dumped waste into it. Coursing now through a postindustrial valley, the river, a much more controlled waterway these days, suffers industrial legacies, despite flowing cleaner, in part, in places, than it did even 10 years ago.[18]

The Lackawanna has borne many insults. In the 1950–1960s, it carried fluorescent light bulbs and volumes of "filthy and smelly upriver water that contained fertilizer runoff, chemicals, dyes and debris." Getting wet in it meant clothes a "color and smell [that] were unmistakable." A former mid-valley resident remembers a meat processing plant in Olyphant dumping "blood and guts from the cattle . . . into the river. The smell was disgusting and you could hear the animals crying. It was not a pleasant place to be either near or to live by." Despite such assaults on their senses, neighbors hardly ever discussed health issues or remediation efforts.[19]

"Lackawanna" as American History

In response to the filth and the smell, adults taught children to avoid the river, a fact that W. S. Merwin describes in "Lackawanna," a poem in his Pulitzer Prize–winning collection *The Carrier of Ladders* (1970). An "obedient child" (12) who was "told to be afraid" (16, 37), the speaker once "shrank from" (13) the river, so much so that when he ran "on girders of [its] bridges" (14), he "never / looked" (19–20) at the water. As an adult, though, he experiences a greater "terror / a truth" (24–25), knowledge of American history, which makes for "a black winter all year" (29). Part of the collection's "American sequence," which traces U.S. expansion from east to west, "Lackawanna" urges us to remember, to look at the past with both eyes.[20]

That the polluted Lackawanna crosses measures of time may have led Merwin to know place as a palimpsest, a layered history whose strata are personal, local, and national. The poet lived in Scranton as a child, and the coal mined from under the river heated homes and fueled early industrialization on the East Coast. An origin in this history, the Lackawanna forks where the river absorbs Roaring Brook, which runs beside the stone furnaces of Lackawanna Iron and Steel, former forger of T-rails for U.S. railroads, symbols of American progress. Just as coal mining transformed land along the Lackawanna, the spread of heavy industries, the genocide of indigenous peoples, and the making of land into farms and cities redid the rest of the continent into "what was found later no one / could recognize" (35–36).[21]

A reminder of this deep history, culm, or waste coal, turns the Lackawanna "black" (9). When the speaker steps into the stream—his "Jordan" (42)—with "both feet" (41), he "wake[s] black to the knees" (38), aware now that as the river "flowed from under / and through the night the dead drifted down [it] / all the dead" (32–34). Marked by this water, he not only experiences the passage of time but also stands as a stay against it. Realizing the dark side of American history—suffering and death—he cannot avoid all he sees. He writes poems to call attention, to make sense; no longer "ashamed / at a distance" (43–44), he reads the world more closely.[22]

Anthracite mining damaged the Lackawanna. Dumping coal waste into its feeders killed fish, and culm washed into them filled several beds, especially during heavy rains, forcing streams to cut new paths. Suffering degrees of harm, in 1916, 38 tributaries contributed to the Lackawanna "culm and mine water from 67 collieries and numerous mine openings." Most often, the river writhed through Scranton as "a black, foul, ropy mass of fluid," but it also flowed white, red, and yellow. The colors varied: clear blue at the headwaters, multihued in the valley, and tinged with the orange of acid as it merged with the Susquehanna, which the Lackawanna transformed into a "black swath."[23]

A Lackawanna tributary made this colorful history legal, here and elsewhere. As the nation industrialized, coal trumped water in court cases centered on the water-energy nexus. Forced to choose between national needs and individual rights, judges often sided with mine owners in cases about water pollution. To do otherwise would have ended mining, which would have slowed industrial expansion. In *Pennsylvania Coal Company v. Sanderson* (1886), "one of the most cited water pollution cases in American law," the Pennsylvania Supreme Court chose to define as natural the flow of water pumped from a mine and allowed to drain via a ditch into a creek. Absolving the coal company of responsibility for downstream effects, *Sanderson* affirmed a "'property right' to pollute." The stream, Meadow Brook, wound through land that would become the Marywood University campus.[24]

The ruling cast a long shadow. State efforts to clean up rivers and streams bracketed off coal mining. In 1905, Pennsylvania passed An Act to Preserve the Purity of the Waters of the State for the Protection of the Public Health but exempted "waters pumped or flowing from coal mines," mainly due to *Sanderson*. The case kept the state water sanitary board from addressing acid mine drainage, and in 1923, Pennsylvania admitted that some streams were so polluted that people might as well keep trashing them, a classification scheme it abandoned in 1944. Meantime, in 1937, the state again chose mining over water when it empowered the sanitary board "to issue and enforce waste treatment orders to all municipalities and most industries but exempted acid drainage," although it did outlaw "gravity drainage of mine water discharges from underground mines." Signaling a legal turn from siding with coal to siding with water, the state government got serious about mine acid only in 1965, at the tail end of the anthracite industry in the Lackawanna Valley.[25]

By then, the water-energy nexus had utterly reshaped the land. Carving out voids beneath the valley, coal companies redirected its natural drainage, creating new ways for water to flow through the underground. Coal-mining, canal-making, and rail-riding outfits, they dammed swamps, straightened streams, and bridged rivers. Early on, they used waterwheels and steam engines in tandem to lift water from deep mines and pull coal cars up graded inclines. Water guided where crews opened pits, laid rails, and built breakers; at the same time, mine water choked brooks with coal, scoured denuded hills, and killed stream life. A recent replumbing of the region, gas-fired electricity generation has transformed former rail lines and colliery sites into pipelines, pumping stations, and power lines. Renewable energy efforts, meanwhile, have made meadows and woods into windmills and potential hydropower ponds.

2

Reaching Water

Five rivers—the Lackawaxen, Lackawanna, Susquehanna, Delaware, and Hudson—determined how Lackawanna anthracite first reached major markets. To find Philadelphia in 1815, trickles of coal followed the Lackawaxen and Delaware Rivers. By 1830, large-scale shipments floated across the Delaware on the way to the Hudson River and New York. Meanwhile, the Susquehanna promised access to upstate New York, the Erie Canal, and the Great Lakes. Acknowledging debts to water, early anthracite operators baptized their corporate selves as the Lackawaxen Coal and Navigation Company, Delaware and Hudson Canal Company (D & H), Lackawanna and Susquehanna Railroad (L & S), and Delaware, Lackawanna, and Western Railroad (D. L. W.).

These companies contrived a water-energy nexus that pulled the Lackawanna Valley away from Philadelphia toward New York. Largely depopulated of Native Americans, owned by Philadelphians, settled by New Englanders, and overseen by Harrisburg politicians, the valley in 1829 answered to New York investors. Virtually overnight, the place became less an isolated, interior part of Pennsylvania than a colony of New Amsterdam. Before a coal shovel could turn a profit, however, water forced entrepreneurs to discover, by trial and error, whether roads, rails, or canals could best get anthracite to the right rivers.

The transplanted Philadelphians who initiated the infrastructure that moved Lackawanna coal held radically different views on why and where to ship it. Owning mines at the northern tip of the valley, Thomas Meredith envisioned a diverse economy of small farmers, tradespeople, and coal operators; his archrivals, Maurice, William, and John Wurts, plotted a family monopoly of the coal trade. In 1829, with the brothers and the D & H to the south sending coal east to New York, Meredith found himself poorly positioned within the watershed. To survive financially, he elaborated plans for rails and a canal to carry anthracite northward to the Susquehanna; at the same time, he relocated to Carbondale to use the D & H rail-canal network. Both moves backfired, but his skirmishes with the Wurtses guided the transformation of the valley from forestland to denuded industrial powerhouse.[1]

A first family of Philadelphia, the Merediths made money on the high seas. Before the Revolution, Reese Meredith, Thomas's grandfather, captained a major shipping business. However, Thomas's father, Samuel, put his inheritance into land, buying tracts in several states. In northeastern Pennsylvania, he owned a 26,000-acre parcel that stretched 20 miles long and 2 miles wide in a swath from west of Waymart, north along Moosic Mountain, to Hines Corners, near Orson. When a land bubble burst in the 1790s, Samuel sank deep into debt. By 1812, he couldn't pay his creditors, and in 1815, he fled to Mount Pleasant Township, where he built a huge home, which he christened Belmont. Unable to sell the oblong—too much of it was too steep to farm, log, or live on—he died land poor in 1817, leaving his son a tab totaling $13,850. In today's money: $273,890.[2]

Early on, Thomas traveled abroad; in 1800–1801, he visited China and India, worlds away from Philadelphia. Back home, he discovered his father unemployed and burdened by bills. After earning a law degree and admittance to the Philadelphia bar in 1805, he trekked to Wilkes-Barre to manage lands for his father, among others, a task that would become his "life's work." Years later, he confessed that he was not entirely happy playing the good son but that abandoning "these family obligations . . . he was told, would be the ruin of his father: 'I could not stand this appeal. I resumed the care of these lands, determined to struggle to the last in their defence.'" He could hardly imagine then that taking up his father's affairs would lead him to epic battles over coal and water.[3]

Meredith lived at Belmont when roads and rivers ruled transportation. Although they amounted to little more than wide ways through woods, the toll roads that intersected at Belmont connected the five rivers. The Newburgh Turnpike (Route 371) opened in 1811 to Cochecton, on the Delaware, and Newburgh, on the Hudson. The North and South Road by 1819 reached Easton, on the Delaware. The Milford and Owego Turnpike, a Delaware-to-Susquehanna route, met the Belmont-Easton Road a dozen miles south of Belmont.

If nothing else, the corners offered Meredith a good place to collect tolls and meet settlers who might purchase land in adjacent Susquehanna County. To pay his debts, he had one thing: location, location, location.[4]

During the War of 1812, he discovered something just as valuable: coal. When a British blockade cut fuel shipments to Philadelphia, city blacksmiths and manufacturers turned to upstate anthracite, skyrocketing the price in 1813–1814 "about 300 percent," an incentive that piqued Meredith's interest. Long before anyone swung a pick in Griswold's Gap—at the center of the family oblong—he must have known of the coal outcrops that a stream there had uncovered. The problem: getting the coal to the Delaware, the best route to the city.[5]

To move anthracite to Philadelphia, Meredith built roads. Through Griswold's Gap, he cut a "coal road" across Moosic Mountain to the North and South Road in 1814, which allowed him to ship anthracite via the Newburgh Turnpike to Cochecton, load it on boats, and send it down the Delaware. In 1817, he and others financed the Lumber and Stone Coal Turnpike to connect Belmont to the Delaware at Stockport. The first load of lumber reportedly crossed this road in 1820, so it's likely that Meredith also shipped coal along the route, a distance of about 30 miles.[6]

Despite indebtedness, in 1818, Meredith bought land west of his mines, much of present-day Forest City north to Herrick Center, near the headwaters of the Lackawanna. By 1827, his mines attracted the attention of the *American Journal of Science*, which reported that "the Belmont mines are on the Lackawanna creek, in the county of Susquehannah [*sic*], seven miles above Carbondale." Meredith may have opened quarries on both sides of the river, "nearly if not quite, upon the top of the mountains. A vein within thirty feet of the top of Moosic mountain . . . [was] four feet thick."[7]

Although Meredith could reach rivers by road, hauling by horse and wagon was expensive. And raising money to build a coal business in those years was difficult for anyone, maybe especially for a debtor so far from Philadelphia. By the mid-1820s, he realized that roads were no route to success. To turn a profit, he had to cut costs, and cost cutting meant finding better ways to reach rivers.[8]

At Belmont, water drains in two directions: east into the Lackawaxen, which empties into the Delaware, and west into the Lackawanna, which flows to the Susquehanna. The drainage matters because the divide knits an early, nation-defining intersection of water and energy. From the high ground of home, Meredith could peer into each watershed. To the south and east, he saw the distance to Philadelphia; to the north and west, he imagined the route to reach home. Although his lands straddled a gap between basins, he couldn't move coal cost-effectively. However, his rivals, the Wurts brothers, found a way: since they too couldn't easily reach rivers, they built one.

Maurice Wurts and his brothers William and John were not from a first family of Philadelphia; they ran a dry-goods store. What brought them to northeastern Pennsylvania is a mystery. William may have wandered into the region, or the brothers may have received a land grant tied to a federal contract during the War of 1812. No matter the reason, they knew about anthracite because, in 1813, Maurice bought some off two wagons from Mill Creek, near Wilkes-Barre. When the British blockade ballooned the price, the Wurtses, like Meredith, saw an opportunity.[9]

The Lackawanna River uncovered coal. While Meredith hacked a coal road, Maurice Wurts learned about outcrops along the river at Ragged Island, present-day Carbondale. Heeding some advice, he began buying land. In 1815, the year that the Merediths fled Philadelphia, he heard that the Lackawanna had also exposed anthracite in Providence, part of present-day Scranton. Wurts started digging there, but at least one owner refused to sell any land.[10]

Water enabled and hampered the Wurtses' attempts to ship coal to Philadelphia. In 1815, a crew piled sleds with Providence coal, labored over Moosic Mountain, and heaved the fuel aboard a boat on Jones Creek, a tributary of the Lackawaxen; "swollen by heavy rains," however, the stream slammed the raft against a rock, breaking it apart, spilling the stones. Still, in limited quantities, at odd times, in wagons and on sleds, subsequent shipments of anthracite wound their way over the mountain to the Lackawaxen, to the Delaware, and on to Philadelphia. It wasn't much, though, and its arrival was never certain.[11]

Shipping from so far added to the Wurtses' trouble, especially after the war ended in 1815. Peace brought boatloads of Virginia soft coal to Philadelphia, and investors grew reluctant to put money into the hard coal trade, despite decent prices. Meanwhile, closer to Philadelphia, the Lehigh Navigation and Coal Company began remaking the Lehigh River to ship anthracite more easily from Summit Hill to the Delaware; crews cleared obstacles, built dams, and loosed "artificial freshets." In 1820, the company sent more than 300 tons to the city. A hundred miles from home, the Wurtses weighed whether they could compete.[12]

With Moosic Mountain overshadowing them, the brothers had to wonder: how do we move tons of coal up and over a 1,000-foot summit, cart it through woods, float it for miles on rough water, and turn a profit in a city whose people barely know how to light the stuff? With paths for roads, power mainly muscle, and coal fairly heavy, water was the best route to Philadelphia, but you had to reach rivers to use them. Like Meredith, the Wurtses tried roads; in 1822, they cut one across Moosic Mountain, but the effort hardly helped.[13]

Early on at Ragged Island, Wurts workmen quarried anthracite along both banks of the Lackawanna. After carving out coal with drills and wedges, they swung sledgehammers to break chunks into market sizes. At work in winter along a front nearly 1,000 feet wide, they chipped at a seam more than 20 feet thick. In the 1823–1824 season, they had 800 tons ready to sled through Rix's Gap to the Lackawaxen, but "unusually mild" weather limited the haul to about 100 tons. Hindering shipments, this warmth melted the snow needed to move the anthracite meant to heat the homes and power the mills that supported the mining and the hauling.[14]

The Lackawanna River not only exposed anthracite but also menaced the men mining it. At first, crews took coal "from the bed of the river, by diverting the river from its channel." To keep water from swamping the quarry, miners jerry-rigged a "rude pumping-apparatus" that river currents powered. When ice threatened to stop it, workers built under it a grate on which they kept a smoldering coal fire. This pump prefigured a positive feedback loop of the fossil fuel age: burning coal melted ice to keep a river turning a pump that removed water from a working mine, which the river threatened to fill, which required mining more coal to melt more ice.[15]

With roads and rivers unreliable, Meredith and the Wurtses competed to dig ditches. Completion of the Erie Canal in 1825 revealed that canals cut freight costs by 90 percent. To reach Manhattan, where at least as early as 1812 a market existed for hard coal, Maurice Wurts imagined a canal to the Hudson River. With access to New York, he could ignore the Philadelphia market, which mine owners nearer the city dominated. Bringing Lackawanna coal to Manhattan, about as far from Carbondale as Philadelphia, the Delaware and Hudson Canal ran up hills, over streams, and across multiple watersheds. After flowing alongside the Lackawaxen, the channel crossed the Delaware, followed the latter's former course through New York, and ended near Rondout, on the Hudson, a distance of 108 miles. Although the canal made for smoother, more predictable travel than did rivers, the ditch suffered ice, floods, and drought. Opening along its full length in October 1828, with the first shipment of anthracite reaching the city in December, the canal carried coal to the Hudson until 1898.[16]

Building the canal required unfettered access to the Lackawaxen River. In 1823, Maurice Wurts worked with his youngest sibling, John, a state senator, to secure a charter that granted Maurice rights to transform the Lackawaxen watershed for his private ends. Assured a monopoly of the river, Maurice and his backers could control the transportation of Lackawanna coal. Pennsylvania allowed them to "enlarge or deepen [the river] in any part or place thereof, in the manner which shall appear to them most convenient for opening, enlarging, changing, making anew, or improving the channel." Improvement meant that Wurts could "form, make, erect and set up any dams, locks or any other device whatsoever." As Thomas Meredith later charged, however, this initial charter said nothing about the *coal trade*.[17]

Two years later, the Delaware and Hudson Canal Company (D & H) formally organized. At about the same time, the Wurtses and the outfit they first founded, the Lackawaxen Coal and Navigation Company, merged with the new organization, which acquired their rights to levy tolls and improve navigation of the Lackawaxen. In 1825, the Albany legislature approved the formation of the D & H in New York, giving the corporation a greenlight to make the Lackawaxen a branch of the Hudson. Signaling Gotham control of the new company, investors elected Philip Hone, D & H booster and later New York mayor, as its first president. In July, construction of the canal began. Vertically integrated at birth, the D & H introduced corporate industrial capitalism to the little-settled Lackawanna Valley. From about this point on, Meredith spent a large part of his life dueling with the D & H monopoly.[18]

While the Wurtses struck east to the Delaware and the Hudson, a cash-strapped Thomas Meredith foresaw a canal stretching north to Great Bend, on the Susquehanna. In an 1824 letter to Jason Torrey, a local surveyor, he explored possible water supplies for his "projected canal [whose] object is to connect the Lackawanna with Starrucca Creek," a tributary of the Susquehanna, which would have allowed him to ship coal to western New York and the Erie Canal. The plan was to extend the ditch to the Lackawaxen and "make a cross cut to the Lackawanna Coal Mines," the Wurts operation. Meredith also sought Torrey's "opinion on the practicability of uniting the waters of Wallenpaupack and Roaring Brook," a move that would link the lower Lackawanna Valley to the Delaware, another way to checkmate the Wurtses' dream of a transportation monopoly.[19]

Openly challenging the Wurtses, Meredith's plan depended on securing enough water at the highest points along the route, with Griswold's Gap a first concern. Meredith wanted to use "numerous lakes," "valley creeks," and water from the Lackawanna to supply locks across Moosic Mountain and in the crosscut to Rix's Gap. His vision much wider than the plans of the D & H, he may have seen possibilities in connecting distant rivers, but his hopes for canals dried up for lack of money and water.

As caught up as he was in an energy transition, Meredith also sensed the tremors of a transportation revolution. In 1825, he turned from canals to gravity railroads, conceiving the Lackawanna and Susquehanna (L & S), a line to connect Great Bend and Pittston, each on the Susquehanna, with rails running north–south through the middle of the Lackawanna Valley, a distance of about 70 miles. He also foresaw an extension through Griswold's Gap and then southward to connect with the D & H canal, which would have allowed him access to the Hudson. His wide-angle view mirrored that of Jacob Cist and other Luzerne County businessmen who advocated a "lakes-to-sea system" of canals radiating from Philadelphia, with one along the Lehigh River to Wilkes-Barre and up the North Branch of the Susquehanna to the Finger Lakes. In comparison, the D & H project appears too narrowly focused, which, of course, may be why it succeeded.[20]

At public meetings in 1825, Meredith exhorted people to back construction of the L & S because plans afoot in New York would link the railway to the Erie Canal. Attendees also rallied around the idea of running the road down the Lackawanna Valley and on to Wilkes-Barre, on the Susquehanna. Meredith described L & S as a company that would "carry on the coal trade extensively," though not exclusively, as the D & H intended. The success of the Erie Canal, though, attracted major investors more to canals than to railroads.[21]

In mid-January 1826, Meredith petitioned the state to incorporate the L & S, with capital stock of $150,000, a request approved in March. Stipulating that the route would run from the Belmont mines to the Susquehanna River at Harmony Township, the charter conferred power not only to "construct railways, build boats, and erect steam engines, and to lay down tract," but also "to hold 2000 acres of coal land, for the purpose of carrying on the coal trade." In April, the state okayed a D & H railroad as a public highway to connect the company's coal mines to its canal.[22]

The simmering conflict between Meredith and the D & H boiled over in late 1829. The trouble erupted when the D & H denied Meredith access to its railroad, which all but denied him access to the canal. When he argued that public investment opened a private project to public use, the D & H answered that they built it, so they controlled it.

Before 1825, few people knew the confluence of Dyberry Creek and the Lackawaxen River. By 1829, however, the forks had morphed from woods to industrial zone. For the D & H, the forks meant survival. At the site, the company transferred coal from rail to river. If the system failed at this junction of natural and artificial waterways, the company would collapse. And it nearly did.

After building a transportation system, expanding mining, and enduring construction setbacks, the D & H needed a financial recharge. But in 1829, chains on the railroad kept snapping, and poor-quality coal hurt the company's reputation. Even worse, just as anthracite began flowing into New York, water-related disasters halted operations. Within four months: on the Lackawaxen, a flood breached the canal; on the Delaware, lumber raftsmen tore up a dam; and on Moosic Mountain, one pricey locomotive failed to steam and a second proved unusable. Piling on the pain, Thomas Meredith conjured a public relations nightmare.[23]

The wildness of water nearly swamped the D & H. The Lackawaxen's flow could not both float logs and fill the canal, forcing the company not only to build dams but also to dig the river a new channel near a place called "the pulpit," or pool pit, a curve in the stream south of Hawley. Water will find its way, however, despite human efforts to move it to more convenient locations. In April, the raging river ripped out a stretch of the canal and swept itself back into its natural course, creating a mess that took until mid-summer to repair. Rebuilding costs hurt, but the company also hemorrhaged damages to dealers who couldn't use the river to move lumber to Philadelphia.[24]

For the Wurtses, the Delaware was at first a road to riches, but building the canal made it a barrier. Instead of running downstream, coal had to cross it, which required calm water, so the D & H constructed a dam. To adhere to its charter, though, the company had to let the river run free enough for rafts of logs. Faced with competing demands, engineers cut into the wall a 100-foot-wide sluice to allow rafts to pass. For loggers, the dam, despite the gate, hampered free use of a public highway.[25]

Timber was big business in the 1820s, and a lot of logs poured down the Delaware. In 1825, the Lackawaxen contributed just over three million feet of white pine, with rafters making nine dollars per thousand feet. The figures don't include logs dumped into the upper Delaware, some of them hauled overland from the Susquehanna or brought to Stockport on the Lumber and Stone Coal Turnpike. In 1830, Meredith claimed that the canal's demands for water hurt raft traffic: "In the fall of 1827, nearly fifty rafts passed down the West and Dyberry branches of the Lackawaxen. Since that time, few have descended that section of the river; and none without injury."[26]

In April 1829, frustrated rafters blew out 80 feet of the Delaware dam. The year before, ice had marred the sluice apron; in response, the D & H had dumped rocks into the breach, which the river, however, simply carried further on, piling them in a way that made the passage of rafts impossible. After the company refused to hear repeated requests to repair the apron properly, upriver lumber dealers took matters into their own hands. Given his interest in the Lumber and Stone Coal Turnpike, Meredith may have been one of them.[27]

Maurice Wurts calmed the situation enough that in December, he convinced 21 lumbermen and rafters to sign a pledge to protect the interests of the D & H. In exchange, the company cleared them of responsibility for destroying the dam. This agreement may be why rafters publicly rebuked Meredith a few weeks later when he aired their grievances. Perhaps not incidentally, the company had arrest powers and could level a "penalty of four times damages for anyone impeding navigation" of coal boats. Presumably, no lumberman enjoyed similar privileges.[28]

Despite its length, the D & H canal stopped 17 miles short of company coal mines. Moosic Mountain loomed in the way. To bridge the gap, engineers imagined not only two series of inclined planes and stationary steam engines to pull coal cars to the summit but also steam locomotives to tug them across downside levels. Just like the canal, stationary steam engines needed a water supply. The location—and success—of the rail-canal system depended on the "greatest elevation at which water could be obtained from the natural ponds strung along the western terminus of the route." Just as Keen's Pond supplied the canal, Panther Falls Creek supplied the stationary steam engines.[29]

The latest technology, steam locomotives carried their own water. In summer 1829, the company bought four in England, among them the *Stourbridge Lion* and the *Pride of Newcastle*, shipped them up the Hudson to Rondout, transferred them to canal boats, and unloaded them at the forks of the Dyberry. When a crew tested the *Pride of Newcastle*, they discovered that it would not burn anthracite. The company kept mum about this expensive embarrassment.[30]

Two weeks later, to great fanfare, the D & H publicly tested the second engine, the *Stourbridge Lion*. At the "bustling construction site," a hemlock rail bridge crossed the Lackawaxen, a half-finished hotel faced the river, and a nearby mill made logs into lumber. On August 8, workers, boosters, and investors thronged the riverbanks to witness the *Lion* hurtle west three miles and back, its whistle resounding in the woods.[31]

Although the engine boasted on its face the image of a predatory king, the most exotic thing about the iron horse was that the machine transformed water into power, a fact some in the crowd refused to credit. As a local newspaper observed, "when it was further announced that by pouring water into this horse's mouth, it would emit from its nostrils clouds of noisy steam, its iron feet would grind on steel rails, and the thing would run: 'I don't believe it!'

was the derisive reply of the sceptical [*sic*]. And when the day came, what a crowd! What anticipation! What mocking laughter!"[32]

The skeptics got it half right. The *Lion* ran on boiling water, but wooden rails and trestles cracked under an engine built too big. With the *Lion* sidelined, the company returned to homegrown horses. But the quick trip, wood fired and steam driven, previewed trips on tracks across the continent. It also staggered the D & H: the disappointing news played havoc with company stocks, endangering the enterprise's existence.[33]

The *Lion*'s run reads as prophetic of the joy and violence that have ever since accompanied fossil fuel extraction. Although largely forgotten, agonized screams pierced the celebration, which included volleys of cannon fire. When one cannon became too hot for a Mr. Griswold to hold the vent, a premature blast caught Alva Adams as he tamped home a shot; the explosion rocketed the ramrod across the river and tore up Adams's arm, which was soon amputated.[34]

With the locomotive trials less than successful, the million-dollar D & H soon settled on a gravity railroad to connect Moosic Mountain to the canal. The force of gravity brought loaded cars downslope, their momentum pulling empties up. At the canal basin, crews stockpiled the coal for shipment to Rondout, on the Hudson; from there, coal boats slipped south to Manhattan. The rails carried coal to a fake river, which floated it to a natural one that brought it to a less-than-natural place, New York.[35]

The lesson: the rise and fall of ancient seas, the location of Carboniferous swamps, and the flow of rivers, streams, and springs defined where and how Meredith and the Wurtses built the infrastructure they needed to move enough coal. Their networks expressed uncertainties that still haunt fossil fuel economies: exhaustible, coal went one way; canal boats and railcars wore out; its own agent, water cycled through the system.

Access to the canal triggered a war of words in 1830 that nearly undid the D & H. On October 9, 1829, the company first used its gravity system to send coal to the canal. Meredith assumed that he could use the line to ship his own coal. After all, D & H president John Bolton had told him, "To yourself, as the owner of several coal beds, and especially of one, which I hear you have lately purchased, near the termination of our rail road, the measure now adopted, must be interesting, as affording you the means of bringing it into early and profitable operation." In an October 24 letter to Bolton, Meredith asked the company to determine the toll to transport anthracite to the head of the canal; he had 150 tons ready for shipment. Receiving no reply, he began advertising in early November for teamsters to haul his anthracite to the forks of the Dyberry. In mid-November, Bolton confirmed in person that company rails would not "carry any coal for individuals this season." Denying Meredith and others access to the railroad all but denied them access to the canal, effectively killing their best chance to profit in Manhattan coal markets. Unlike other canals, the Wurts ditch had one purpose: to float D & H coal to New York. In denying others use of the line, the company made clear that it would rule Lackawanna anthracite.[36]

A water-energy nexus is a concrete and abstract construction. As much as it produces physical power, it also circulates power in words, which shape and reinforce perspectives through which people know the world. Written texts guide views of reality and influence the choices that people make, whether political, social, or economic. Raging in writing, Meredith responded to Bolton in a 30-page pamphlet, completed and published within a month, a document that expresses still-unresolved tensions at the heart of corporate capitalism. The rhetorical explosion endangered the company's existence, delayed its expansion, and defined how local infrastructure developed.

Like many of his contemporaries, Meredith assumed that building roads and digging ditches first served the public interest, not private profit. *Monopoly Is Tyranny* demanded that the D & H open its railroad and canal to the coal of other colliers. Meredith accused Maurice Wurts not only of changing maps, redirecting streams, and flattening mountains but also of keeping key facts from public view to secure "an enormous rate of toll on articles of the first necessity." According to Meredith, the public knew nothing of the Wurtses' plans until the appearance of a pamphlet in 1824. This publication asserted that Pennsylvania had given the D & H control of coal in the Lackawanna Valley, despite not only a founding charter silent about anthracite but also state approval of the gravity railroad as a "*PUBLICK HIGHWAY*!"[37]

Meredith branded the D & H a monopoly, noting that the company would not even allow goods on empty railcars returning to its mines, nor would it permit use of its railroad to transport lumber to the canal basin. Evidence in support of the latter charge emerged a year later. In response to an 1831 letter, the company flatly refused to transport lumber on the canal, claiming that it would only haul "coal and other articles for their own purposes." Coming on the heels of Meredith's polemic, the request "revived in stark terms the question of whether the D & H was a public highway." Although the company soon backed away from its refusal, allowing lumber on the canal at "favorable rates," the lumber dealer needed to get his product to Honesdale. Not until 1843 did the gravity railroad handle anything but coal. By keeping tolls high and refusing to open an outlet lock to the Delaware, the D & H kept the canal a limited-access waterway to the Hudson.[38]

Claiming to speak for independent miners, Meredith demanded equal access to transportation infrastructure, assuming one could afford the tolls and haulage rates. He believed that equal access would establish a large number of independent collieries, small farms, and single-family businesses in the valley. In response, the D & H asserted a right to deny others the use of its works, a position meant to ensure that it would find its financial footing and endure as a vertically integrated corporation. Seeing no problem with monopolizing anthracite, the D & H understood that controlling access to coal required controlling access to the canal. For a decade, though, Meredith's drumbeat about monopoly made the D & H hesitant to extend its reach further down the valley.[39]

Some locals saw no downside to a D & H monopoly. At 1830 meetings in Susquehanna and Wayne Counties, attendees railed against Meredith, arguing that businesses already depended on the canal and that many people new to the region owed their jobs to the D & H. Perceiving the company as vital to the region's interests, many either rejected the idea that it was a monopoly or didn't care. Unhappy with Meredith, Wayne County residents wondered, "Shall we permit ourselves by one mighty blow to be struck into that dark abyss of a confined interior, from which we have just emerged, without exerting every nerve to prevent it [?]" If the D & H folded, some reasoned, Lackawanna coal would go unmined, the canal would go unused, and local industries that relied on mining or the canal would never materialize. At the dawn of the fossil fuel era here, coal meant progress and prosperity.[40]

Unlike anthracite or timber, water returns as itself to the commons. Hard coal burns to carbon and lumber never becomes the tree again, but water cascading from waterwheels or rising as vapor from steam engines is still water. The D & H needed permission from the state to use for its private ends a public resource—water—that neither respects political and economic boundaries nor transforms easily into a private good. Drawing on rivers and streams to keep its boats afloat, the company may have dug the ditch, but the commons supplied its motive medium. A canal is a ditch and its water. If one argues that the D & H had the right to limit access to what it built, isn't a dry ditch just a high-sided road and not the canal that the state approved? The question behind the question: can water be privatized?

Meredith's story echoes today because the U.S. still wrestles with relationships between private profits and public benefits. Communities fund incubators and tax-free zones to compete for businesses, taxes build stadiums to entice teams, and banks get bailed out because they're too big to fail. Meanwhile, governments condemn land to lay pipelines and the Supreme Court okays a city to take one person's property for another's development. All this may lead to jobs and their economic ripples, but the concentration of wealth these days suggests that much of the good flows to the already rich, a fact Meredith experienced in 1830. When he condemned monopoly, no one heeded his warning about the few amassing the necessities of life.[41]

In its dispute with Meredith, the D & H had the last word. When the sheriff moved to sell Meredith Cottage in 1863, the company joined the suit. A public notice described the mix of farm, mine, and mill that Meredith had hoped to see dominate the Lackawanna Valley. He left behind 500 acres, "about Two Hundred Acres thereof improved, with a MILL RACE and POND, with an old SAW-MILL DAM, two old MILL HOUSES, one FRAME TENANT HOUSE, one frame FARM HOUSE, a frame BARN and SHED, a small ORCHARD, three small COAL OPENINGS, and about three hundred feet in length of Railroad from the coal openings to the dumping ground thereon."[42]

Adding insult to injury, in 1870, the D & H invited its rival, the Erie Railroad, to lay tracks from Carbondale to Susquehanna, a town on the river. Generally following Meredith's proposed route for the Lackawanna and Susquehanna Railroad, the Jefferson line passed through Forest City, near his former mines, and Stillwater Gap, to Union Dale, across Ararat, and on to the Susquehanna, opening western New York to D & H coal, a good portion of it mined in Forest City, former Meredith land. In 1872, the company connected this line to Nineveh, New York, naming the route the Lackawanna and Susquehanna Railroad.[43]

The D & H monopoly did not last long after Meredith's death in 1855. Henry Drinker, a Philadelphia land speculator who sometimes collaborated with Meredith, planted the seed that ended D & H control before it began. In 1819, Drinker foresaw a gravity railroad stretching from Pittston to Roaring Brook to Delaware Water Gap, an idea Meredith revisited in his letter to Jason Torrey. Surveyed in 1824, the never-realized system was to "run with a view of inclined planes operated by water, and perhaps a canal over the more level portion of the way." In 1856, the Delaware, Lackawanna, and Western Railroad developed the route as a locomotive line, ending the D & H stranglehold on transportation of anthracite out of the Lackawanna Valley. Later the Anthracite Road of Phoebe Snow, the D. L. W. carried coal in two directions: east through Cobb's Gap to Delaware Water Gap and north through Leggett's Gap to Great Bend on the Susquehanna. Meredith's suspicions of the D & H raised an issue that the nation didn't settle until a Supreme Court decision in 1920: railroads that owned coal mines constituted a monopoly.[44]

To access and control water, Thomas Meredith and the Wurts brothers built the infrastructure that spawned the collieries, railroads, and related industries that soon dominated the valley. Only remnants remain: artificial lakes, earthen dams, and drainage tunnels mix with rusted rails, sealed shafts, and abandoned equipment. Some things have been repurposed. In Scranton, the D. L. W. Railroad station is a hotel; the D & H headquarters in Honesdale houses the Wayne County Historical Society; the last deep mine to operate in Lackawanna County attracts tourists. The rest was razed or cannibalized, dismantled and sold for scrap, or simply left and forgotten. The canal is gone, local rail is a phantom of its former self, and people wonder at what survives. Meanwhile, as a new regional water-energy nexus takes shape, the five rivers flow on . . .

3

Mining Water

Myths remember. In the beginning was nothing, an emptiness the Norse named Ginnungagap. Then within the void appeared a well, the source of 11 rivers that flowed on in the North until they hardened. Rain added layer upon layer, ice filling the gap until it touched warm southern winds. Melting, frost dripped into the likeness of a man, Ymir, an ogre. From parts of his body, the world was made.[1]

Beneath the world tree murmurs a spring of wisdom. When Odin, the chief god, seeks a sip, Mimir demands an eye in exchange. After the god hands it over, drinks, and waits, the rememberer tells him that to be wise is to see with both eyes. The lesson: more than a thirst for more, wisdom requires depth perception; to see with one eye is to only half discern. If humans are godlike, as some think, we are like Norse gods: powerful but not all-powerful. Wising up might mean being mortals who know their limits.[2]

Technological change triggers forgetfulness. With so many cell phones around, fewer and fewer folks recall rotary ones, which are not ancient devices. Although, in the nineteenth century, waterwheels powered industry in the Lackawanna Valley, even as early as 1902, local residents sometimes forgot this fact. In January of that year, a crew in Carbondale at work replacing a bridge with coal storage discovered a wheel buried below river level. For a week or so, people puzzled over it, this relic of another age, despite its discovery within living memory. Settling the matter, a longtime D & H employee confirmed that three wheels once turned there. One pulled loaded cars from a coal mine; the others hauled them up plane No. 1 of the gravity railroad.[3]

Faith in our power too often tempts us to forget the complexity of nature, which we replace with simplicity, complicating—and compromising—water cycles. To control water in South Florida, in 1962–1971, the U.S. Army Corps of Engineers remade the winding Kissimmee River into a ditch. Crews constructed "1,000 miles of canals, 720 miles of levees, sixteen pump stations, two hundred gates, and other instruments of order . . . meant to calibrate the rain machine to human time and space." The project played havoc with water and wildlife and triggered severe flooding and overdrawn aquifers. To reclaim the river, in 1992, crews began "dynamiting water control structures, remaking meanders and oxbows, and reestablishing 12,000 acres of wetlands." Another attempt at river management, the restoration tried to undo artificial disruptions of natural systems with new systems that mimic natural ones. Even when we hold on loosely, we won't let go.[4]

Dewatering

On the surface, coal mining wrote Scranton's story; at deeper levels, however, water shaped the city. While anthracite operators altered streams, drained swamps, and raised reservoirs, miners not only manhandled steam-driven equipment but also dealt daily with make water, or natural seepage. Messing with the movement of groundwater, mining accelerated acid mine drainage, a natural process that mining made unnatural.[5]

If you dig down far enough, you discover that water drowns your digging. The deeper you go, the wetter you get. Wrestling with a shape-shifter, miners in the Lackawanna Valley could not simply gather water, pile it aside, and be done with it. Dewatering was a 24/7 job that required moving water up, out, and away from mines, which meant directing it into streams and rivers. Coal drove pumps, but when pumping costs rose and demand for anthracite slackened, companies couldn't afford to keep tunnels dry. Miners may have first known water as a nuisance and then experienced it as a threat, but finally it overwhelmed them. In the immediate aftermath, the subsidence narrative became starkly visible.

Mining coal in the valley offers lessons about working in a wet world. Water never stops circulating, so responding to its movement, a constant challenge, created a self-reinforcing loop that irked owners, who had to keep enhancing infrastructure to keep ahead of it. To cover costs of dewatering, they dug deeper, hired more workers, and mined more coal, which meant mining more water. Pumping both burned coal and enabled its burning, and each firing cast more carbon into the atmosphere, which today unleashes more rain locally. In interfering with the flow of water through the watershed, miners flooded the valley first with acid mine drainage and then, now, with the effects of climate change, which include erosion, flooding, and, not so distantly, ocean acidification.[6]

Although water seeped into mines and followed men through mine openings, streams sometimes threatened underground workings. In 1917, a subsurface rockslide in Scranton cracked the bed of Stafford Meadow Brook, endangering the National mine. Near Forest City, fearful engineers at the Clinton colliery rerouted Clarks Creek to fend off flooding. Other companies built "miles of wooden flumes" to steer flows away from shafts; in 1913, a conduit near Carbondale carried Lees Creek to the Lackawanna. Some collieries erased streams entirely. Maps from the 1890s include creeks not found today, although their locations still flood during heavy rains.[7]

Dewatering costs contributed to killing coal mining in the Lackawanna Valley. Mining water was not the only cause, maybe even not the most important; market forces, changing consumer demands, and advancing technology played parts, but ultimately water drowned it, slowly, prompting a flight of people and businesses from the region in the 1950s. Ever more complex and powerful solutions could not dodge a key fact: the deeper you mine, the more water you bail. Eventually, mining doesn't pay; the slide to collapse quickens and end days dawn. Shifting to another fuel doesn't help, because wrestling with water haunts not only coal mining but also renewables, nuclear power, and the extraction of oil and natural gas.[8]

To mine water was to burn coal. In the early twentieth century, collieries used 8–10 percent of mined coal to power steam engines, mainly to pump water. Some operations ate up to 40 percent of coal output to run engines, so much that "operators often wish[ed] that they might sell the water and run the coal to waste." Adding to costs, companies worked smaller pumps for shallower depths and kept excess capacity to cope with emergencies. Installing a pumping plant in 1911 took between $10,000 and $50,000, so recouping investment meant that "a great many thousand tons of coal must be mined," all at no profit.[9]

Ever more efficient and powerful pumps did not save the industry. Over a 100-year period, hand devices gave way to "mammoth steam and electric motor driven pumps of enormous capacity." In 1899, the D & H operated in the Von Storch mine a Jeansville steam pump capable of lifting 3,200 gallons a minute. By 1932, electric machinery at the Jermyn colliery of the Hudson Coal Company pumped 33,000 gallons per minute, or 47.5 million per day. Still, improved technology couldn't keep up.[10]

Collieries mined more water than coal. In 1905, anthracite operations pumped 633,000,000 gallons per day, enough for "a lake a mile long, one quarter of a mile wide and nearly 20 feet in depth." In 1900–1920, miners removed about 10 tons of water per ton of coal. In 1931, the ratio jumped to 12:1; just six years later, in 1937, it reached 18:1. Although, by 1944–1948, the average for all four anthracite fields hit 19:1, in the northern field, companies mined 21 tons of water per ton of coal. Nearly unsustainable, in 1951, it spiked to 32:1 and soon went higher, to 40 tons, a yearly burden on the industry that topped $12 million. In 1957, the ratio ran to 56:1, outstripping profits. Without a solution, engineers warned, the industry would collapse.[11]

Defining all change, the laws of thermodynamics stand behind the mine water problem. Discovered, in part, by people like James Watt tinkering with steam engines, the first two laws state that "(1) the energy of the universe is constant; (2) the entropy of the universe tends toward a maximum." From the Greek for "transform" or "turn," the term "entropy" names a measure of disorder in a system. As heat in a steam engine flows from hot to cold, the transformation causes the engine to work. But as this happens, energy—the capacity to do work—becomes less and less available; steam cools to water, which must be reheated to do more work. Without added heat—in the form of, say, burning coal—the system moves to "a lukewarm equilibrium from which no further useful work can be derived." Physicists call this heat death. Suggesting that the universe hurtles toward exhaustion, the second law undermines faith in limitless progress. No one, and nothing, lasts forever.[12]

The machine that James Watt improved dewatered mines. In 1712, early in industrialization, the Newcomen engine, a coal-powered contraption, pumped water from a mine in England. In the 1760s, Watt reimagined Newcomen's atmospheric design, and in 1776, he put a steam version to work draining a mine and blowing bellows at an iron foundry. The beating heart of any deep mining operation, a steam-powered pump burned coal to boil water to raise water from workings. The cycle soon intensified: the engines ate more coal and boiled more water as they lifted ever more water from ever deeper mines.[13]

Just as a steam engine captures the flow of heat from boiler to condenser, a waterwheel captures the energy of water flowing from a high point to a lower one. Redirecting a stream and returning it immediately to nature, a wheel can intervene but not transform. A pipe-fed steam engine, however, transfigures water, increasing the capacity of humans to reshape their surroundings. In the transition to coal, water remained a constant, but the move from one water-energy nexus to another was not a smooth, abrupt swapping of waterwheels for steam engines.[14]

Long into the nineteenth century, waterwheels competed with steam. In 1842, 50 wheels tapped a river 12 miles away to dewater mines in Cornwall.

Beginning operation in 1854, Lady Isabella, one of the world's largest waterwheels, 72 feet high and 6 feet wide, drained mines on the Isle of Man and ran almost continuously until 1929. Watt used wheels to manufacture steam engines, and in 1827, a waterworks that rivaled his invention began operation in his birthplace, Greenock, Scotland.[15]

Lackawanna Valley mines used both, sometimes in tandem. With early stationary steam engines expensive and sometimes undependable, the D & H initially relied on waterwheels, which were cheaper, safer, and sometimes more powerful. In 1837, new wheel-powered pumps drained two million tons of company coal. For 30 years, until 1868, they dewatered D & H mines, raised coal cars, and tugged them up inclined planes. As the gawkers in Carbondale knew, the presence of wheels in mining history makes energy production in the valley uncanny, known and knowable, but not.[16]

The water-to-steam transition first transformed the hollow that would become Scranton into an iron-making site. In 1841, crews completed a waterwheel to run a blast furnace. But as much as water enabled iron making, it also threatened it: low water in Roaring Brook helped to snuff out an initial lighting of the smelter. By 1844, a 90-horsepower wheel powered a rolling and puddling mill; a nearby 40-horsepower one ran a nail factory. Only in 1847 did the Scrantons install a stationary steam engine, which, at 80 horsepower, drove the rolling mill. A waterwheel worked the foundry bellows until 1853, when steam engines replaced it.[17]

The move from water to steam in the Lackawanna Valley contributed to a similar alteration of worksites nationally. In the U.S., steam did not outstrip waterwheels in total horsepower until 1870, and as late as 1900, water powered 20 percent of U.S. enterprises. A ballooning concentration of industries in cities between 1870 and 1880 spiked steam use by 80 percent versus 8 percent for water. Sealing its dominance, between 1869 and 1889, steam accounted for 96 percent of increased horsepower in U.S. industries. Running steam engines here, miners dug deeper, railroads multiplied, and textile makers expanded.[18]

To hold off rising water, anthracite collieries used big buckets, compound steam pumps, and electric turbines. Engineers at deep mines often placed hoisting canisters in dedicated shafts and ran them from surface air pumps. A vacuum vessel could raise 8,000–10,000 gallons per hour, but a Galloway lift, which fit inside a fixed tank and filled automatically, could raise 20,000–25,000 gallons each hour. Unlike other systems, hoisting avoided underground floods, mine subsidence, and high repair costs.[19]

A typical steam pump moved water 600–700 feet, so deploying them in deep mines often required more than one lift. Using condensing engines depended on the temperature and acidity of the mine water: the hotter and more acidic, the sooner pump parts would corrode. Some operators lined pipes with wood swelled with fresh water because mine acid could eat the iron from a shovel overnight. Unlike pipes and pumps, however, nothing protected boilers from corrosion.[20]

To enhance longevity and increase volume, after 1905, hard coal operators installed turbine pumps, which were submersible and could operate at any angle. Their design kept bad water from fouling bearings; the copper-and-tin casing of one D & H pump withstood the ravages of acid for at least two years. In 1907, the D. L. W. put to work an electrically driven centrifugal pump to remove 5,000 gallons per minute from a Scranton mine.[21]

Corrosion drove up costs because acid quickly ate brass, which sometimes lasted a few weeks or sometimes only hours. In operations that relied on moving highly acidic water, labor and parts bills often became "exorbitant." In response, collieries chemically treated water to "protect pipe, pumps, valves, tanks, screens, and the metallic lining of chutes." In 1904, miners added soda ash, which made an acidic sludge that further polluted streams, and in 1932, operators put lime in mine water to prevent problems at washeries. Chemical mitigation efforts cost companies dearly, though, about "fifty to ninety cents per ton of water."[22]

Collieries fought rising water in other ways. Engineers not only designed some mines to drain by gravity but also drove tunnels to draw off others. Completed in 1895, a drought year, the Jeddo Tunnel emptied more than 32 square miles of mines and collected over 50 percent of this surface area's precipitation. About 9,000 feet long and driven in 1889, the Centralia tunnel siphoned water from several collieries near Ashland. In the southern anthracite region in 1907, the Lehigh Coal and Navigation Company drained mines through a tunnel to the Lehigh River, a distance of 12 miles. Saving the expense of "big pumps, engines and men to run them," in 1906, the Pennsylvania Coal Company drove a tunnel just over a mile long under Marywood to empty acid directly into the Lackawanna River. To reduce pumping costs, the company also completed the 7,900-foot Pittston Water Tunnel in 1911, which carried mine water to the Susquehanna.[23]

Coping with corrosive water was a necessity for miners but not a requirement for landowners. Engineers designed dewatering systems to move water away from mines, but property owners had no obligation to do anything about water flowing "naturally" from abandoned or, in some cases, active mines. No one held landowners accountable for the environmental costs of acid, and many were reluctant to backfill stripping pits in hopes of someday remining them. Behind this inertia stood an out-of-sight, out-of-mind attitude that has long plagued local land and water.[24]

Conowingo Tunnel

In the 1950s, mine water needed a fresh look because the problem was pushing the anthracite water-energy nexus to a breaking point. Seeing only a design problem, federal engineers proposed a silver-bullet solution: the Conowingo tunnel. Artificially directing a natural flow made unnatural, the tunnel not only would have diverted 80 percent of anthracite mine drainage but also would have driven a stake into the heart of Chesapeake Bay, an outcome few foresaw, despite evidence of mine acid toxicity. Unbuilt, the tunnel represents an important example of the "away" mentality of managing pollution.[25]

By the 1940s, water threatened to destroy the hard coal industry. In 1941, state officials named economic trouble in the Anthracite Region the biggest social problem in Pennsylvania and asserted that solving it meant ending the mine water problem. In 1944, dewatering had become so nettlesome that the U.S. Bureau of Mines directed engineers to study it. A decade later, in 1954, researchers recommended an audacious plan: drain most anthracite mines using a tunnel that would begin at the Eddy Creek colliery in Throop, cut through the Lackawanna and Wyoming Valleys, and strike straight south underground to Conowingo, Maryland. The gravity-fed tunnel, 137 miles long, was to empty via a concrete canal into the Susquehanna 10 miles from the river's mouth. A proposed infrastructure project to protect existing infrastructure, the tunnel echoes similar efforts today, such as the billions invested to save Houston's oil industry from flooding.[26]

Planners designed the tunnel to prolong anthracite production. High operating costs made hard coal less and less a winner against oil and gas, so dealing with dewatering would enhance its competitiveness. Noting that the U.S. had become a petroleum importer in 1947, tunnel designers also invoked World War II as a reason to keep the industry alive: if oil imports ended, the nation would need more domestic coal. In 1957, officials pegged the life of the coal industry in the Lackawanna basin at 10–15 years; a drainage system might extend it to 33. Although mining in the Wyoming basin might last 75 more years, solving the mine-water problem would stretch its life to 100. The southern field was less of a problem; it was estimated to last 420 years. Across the region, enough coal remained to mine for two centuries.[27]

Engineers urged that the tunnel begin in the northern field because water affected it more than it did other anthracite areas. The field's canoe shape ensured that rainfall collected in the coal measures beneath the Lackawanna and Susquehanna Rivers. In 1948, the Lackawanna Valley shed 135 billion gallons of runoff. Of the 45 billion gallons pumped from valley mines, about 10 billion, or 22 percent, came from "direct surface seepage," about the same amount entered through 52 streambeds, and the rest, about 25 billion, percolated through the channel of the Lackawanna River. Wyoming Valley mines pumped 21 percent more water than Lackawanna mines. In total, 30 percent came from direct seepage, 21 percent from 59 streambeds, and 49 percent through the bed of the Susquehanna. Within the Anthracite Region, the northern field as a whole pumped the most water per ton of coal—23—and it stood to lose the most production due to mine water issues.[28]

Underground Pools

Water is its own thing, poet W. S. Merwin reminds me, but people want it to mean something more. In "The Well," "immortal" water "waits / with all its songs inside it" to "sing again" (2–4, 6). Here time is not history, a record of human striving, but the near-eternal present of the earth in its rotations, "days [that] / walk across the stone in heaven" (7–8). Although in "For Once, Then Something," Frost's speaker may have glimpsed in water a higher reality, in "The Well" one sees only its surface, "the stone sky" (1). When "Echoes come in like swallows" (12), the water "answers without moving / but in echoes / not in its voice" (14–16). This series of sounds says not "what it is" (17) but "only where" (18), in "a city to which many travelers / came with clear minds" (19–20). These pilgrims have "left everything even / heaven" (21–22) and, presumably, Frost's "something," to "sit in the dark praying as one silence / for the resurrection" (23–24), for a welling of water, a flow that would mean a renewal of history, ushering in change, memory, and the old ambiguities of human life.[29]

In ancient Greece, some pilgrims sought advice from the subterranean oracle of Trophonius, in Boeotia, which was not easy to consult. Approaching the god required days of purification followed by sips from two springs: the water of Lethe to help the petitioner forget the past; the water of Memory to help him recall what he would soon hear. After he drinks, he descends into a chasm. At the bottom, a voice tells his future, he falls senseless, and he soon finds himself in the Chair of Memory to repeat what he just heard. Still stunned, the supplicant then regains the surface, reborn, perhaps wiser.[30]

The Greeks also believed that all water, not just spring water, rose from underground reservoirs, an idea that held sway into the eighteenth century. Aristotle identified these pools with Tartarus, the underworld where Zeus imprisoned the Titans, the deposed rulers of a former world. The place where

Tantalus relives his punishment, Tartarus stretches beneath the Lackawanna Valley as a series of acid lakes that recall a more recent former world.[31]

As miners carved catacombs under the Lackawanna Valley, groundwater trickled across coal faces, ran along gangways, and collected in sumps. Pumps and drainage tunnels drew the water to the surface, in the 1940s and 1950s on average as much as 43 billion gallons yearly, but the end of deep mining slowed this hybrid hydrologic cycle. When pumps stopped in 1961, 160 billion gallons pooled below in two major basins, leaving invisible lakes that shadow surface ones. A pool stretching from the Archbald pothole to the Old Forge borehole holds 130 billion gallons, almost twice the amount in Lake Wallenpaupack; the other, beneath nearby Carbondale, contains 30 billion gallons. The mine pools of the entire northern field top 435 billion gallons, about the size of Cayuga Lake. Dubbed the Acid Mine Lake, these groundwater cisterns contribute "naturally" to surface waters.[32]

Mine pools rise and fall with the weather just as surface pools do. Ninety feet below our house, the shoreline of the pool under Scranton expands and contracts, which, over time, weakens pillars, which can cause surface subsidence, which often allows more water to pass below. A typical rainfall enters mines, flows through, and resurfaces in Old Forge within 24 hours. Acid draining from a mine pool also varies by season: more in spring, less in fall.[33]

Mine pools mirror surface pools structurally. The pools have two layers, top and bottom. Through the top layer circulates surface and groundwater seepage; the less well-circulating bottom layer collects "relatively high concentrations of acidity and metals." Stagnant and toxic, the bottom moves most when flushed by large volumes of rain. Surface waters in temperate regions have layers, cold and warm, but, unlike in mine pools, each fall and spring they mix, an overturning that renews.[34]

Pumping costs created another feedback loop: as collieries abandoned mines, water rising in empty tunnels threatened active ones, which increased pumping, which raised costs, which put collieries out of business, which emptied more mines, which created more mine pools. By 1957, the ability of active operations to shoulder the pumping loads of abandoned mines was "*rapidly approaching a physical impossibility.*" The loop would soon snap.[35]

The loop determined that the Conowingo project would begin in the Lackawanna Valley. The valley not only had a shorter mining lifespan, but its surface and subsurface waters drained into the Wyoming Valley. If not siphoned off, mine water from the Lackawanna basin would soon spill into the Wyoming basin, dumping the pumping costs of the former onto operators in the latter. Already moving too much water per ton of coal in 1957, Lackawanna Valley collieries faced uncertain futures. To keep ahead of water, they ran 96 pumps, 13 percent of the total used in the Anthracite Region; coping with the buried valley of the Susquehanna, the Wyoming basin operated 343 pumps, 48 percent of the total. Keeping operations in the Wyoming basin above water meant draining Lackawanna mines.[36]

To prevent underground floods, the Pennsylvania coal mining law of 1891 demanded that companies leave barrier pillars between workings. Serving as boundary markers and water-control walls, the barriers defined underground mine pools. By 1952, about 106 billion gallons had accumulated in 150 pools. As a point of comparison, Lake Wallenpaupack, which covers 24 square miles, contains 71 billion gallons. In the Wyoming basin in 1957, barriers separated 60 mines, stretched for 171 miles, and held 21 billion gallons; as coal companies closed in 1948–1957, the number of pools jumped from 19 to 34. In the Lackawanna basin, barrier pillars ran for 87 miles, confined 21 pools, and affected 48 mines. If Lackawanna Valley pools poured into Wyoming Valley operations, barrier pillars, already compromised in many places, might fail, flooding mines and making it physically and financially impossible to continue mining.[37]

History made matters worse. Early operators in the northern field did not value barrier pillars and did not plan much for mine drainage, especially complicating later efforts to deal with the buried valley of the Susquehanna. As pumping costs rose in the first decades of the twentieth century, collieries turned to strip mining, which allowed water another way into deep mines. As early as 1913, stripping was widespread around Hazleton, and by 1951, aboveground operations accounted for about a third of all anthracite mined. By about 1960, surface mining outstripped subsurface production. Hundreds of abandoned bootleg mines dug in the 1930s and 1940s—whose locations and water levels no one knew—added more uncertainty.[38]

Old Forge Borehole

Water remembers on large and small scales. Across millennia, ice tracks atmospheric oxygen levels, ocean sediments pile fossilized lives, and rising seas redraw shores in a wetter world. Just as a pond remembers glacial gouging, so a stream in flood finds former courses and a puddle recalls recent rains. Forgetting this, we hurt ourselves and the world.[39]

Toxic groundwater reserves, acid mine pools produced poisoned springs, or outfalls of acid, a nature-culture lesson of the extraction narrative. As it surfaced, the water reminded locals of the underground voids that others created, abandoned, and forgot. Made visible, these once-suppressed memories clouded and killed living waters. By the 1920s, acid made lifeless much of the Lackawanna River, a fact discounted for decades. Water, however, wouldn't let the region—or the nation—forget that repressed traumas wrote its "cheerful modernist narrative of progress, consumption, and growth."[40]

In the Lackawanna Valley, mining ended, but not. After the last pumps stopped in 1961, owners closed collieries, mines flooded, and acid found new ways into rivers and streams. Abandoned surface lands could be forgotten, many assumed, but underground infrastructure still demanded attention. In Old Forge, Duryea, and upper Pittston, water rose from below, threatening to flood homes, forcing the state to act. To siphon water from beneath valley towns, in 1962, crews drilled a borehole on land of the Pennsylvania Coal Company to tap the "Middle Red Ash seam of the abandoned Old Forge Colliery." Keeping mine water from flooding homes and businesses, its outflow now maintains a steady level in the pool that stretches to Archbald. But the Old Forge borehole concentrated rather than solved the mine water problem. An artificial spring that runs naturally, by gravity, the borehole still spews acid, making visible environmental costs that collieries kept off the books, enabling past corruption to corrode the present.[41]

Another example of an "away" philosophy of dealing with an environmental problem, the borehole directs damage downstream; in effect, the state decided that the Susquehanna River could clean up what the Lackawanna could not. Gushing 37,000 gallons per minute, or over 130 million gallons per day, the borehole reminds residents that local lands and waters still pay costs of coal industry practices. Below the borehole, from Old Forge to the Lackawanna's confluence with the Susquehanna, no "fishery or any significant aquatic community" survives. These last three miles of the Lackawanna, stained bright orange, make the river the "largest point source of pollution in Chesapeake Bay."[42]

No Mimir's well, the Old Forge borehole messes with memory by diverting water from reaching its former levels; unable to find where it seeks to be, the water runs where it wouldn't go, carrying with it rust-colored spots of time, a sediment that sinks and kills. An artificial spring, the borehole pours forth from mine workings our ignorance, not inspiration, let alone wisdom. Telling stories of a bleeding earth, the rust damages waters near and far.

A proposed solution to an environmental problem that was meant to solve another environmental problem, plans for cleaning water from the borehole call for a settling system to harvest the rust for industrial uses. Artificial solutions, however, assume order and stability—projected long into the future—the very ideas that the end of the industry upset and that climate change threatens to undo further. The question that we've only half answered with the borehole is, how do we live with a new hydrologic cycle?[43]

The Old Forge borehole is not the only artificial spring made locally to forestall flooding. In 2013, the Pennsylvania Department of Environmental Protection drilled five holes in Wilkes-Barre along the Sans Souci Parkway to keep water out of other basements and backyards. The state again opened the earth to save the surface from rust let loose by a mining industry that had walked away from responsibility for its effects.[44]

Responding to economic and engineering challenges posed by mine water, Conowingo planners paid little heed to environmental impacts. They saw no problem with concentrating acid outfall into the Susquehanna River, planned no treatment of it at the end of the tunnel, and minimized ecological threats in noting that any problems "must be given careful study." Perhaps in years to come, they wondered, treated mine water might be used in industry or agriculture "in lieu of 'utilizing desalinated seawater.'" Besides, they pointed out, just as the alkalinity of the Susquehanna River neutralized the acidity of water from the Lackawanna, so it would neutralize the flow from the tunnel, which led them to downplay the effects of acid on fish and wildlife below the Conowingo Dam. But if studies ever confirmed that discharging mine water at Conowingo would do more harm than good, the planners simply asserted that an alternate route could empty the acid into the Delaware below Philadelphia, where it "would have a beneficial instead of a detrimental effect" on water "now polluted by industrial wastes." In the 1940s, the average pH of acid mine drainage in the Anthracite Region varied between 3.0 and 3.2. Later analysis pegged the Conowingo discharge at 5.3.[45]

No matter the pH, the mine water problem in Pennsylvania would also have become a river problem in Maryland. The tunnel would have removed over 250 million gallons of water per day from the Susquehanna's flow above the Conowingo Dam, affecting its power plant, which in dry times drew on stored water to the extent that no water flowed below the dam. Greater demands for energy due to population growth and increased frequency of drought due to climate change would have by now cut river flow even more.[46]

Too expensive and time consuming, the initial tunnel project was shelved in 1954. Instead, Congress and Pennsylvania dealt with the drainage problem through the federal-state Anthracite Mine Water Control Program (AMWCP). AMWCP called for deep-well pumps to control water levels in abandoned mines, streambed liners to reduce seepage, relocation of brooks to solid ground, and flumes or pipes to carry creeks away from broken land. These substitutes for the tunnel—like plans for the tunnel itself—assumed an easy sustainability, but even the least expensive options accomplished nothing. Coal companies had the responsibility to implement the initiatives, but most went bankrupt before doing much. With no improvements maintained, diversion disappeared. By 2010, runoff in Nanticoke had broken up concrete poured in 1958 to line Leuders Creek. Lackawanna and Wyoming Valley streams still disappear into mines, and every hard rain pours acid from tunnels and boreholes across the basin.[47]

In 1961, acid mine drainage did serious damage to the Susquehanna. Glen Alden Coal Company, formerly a division of the D. L. W., used AMWCP pumps to dewater its No. 5 mine in South Wilkes-Barre. In October, pumps poured 19 million gallons of acid daily into Solomon's Run, a river tributary. Poisoning the Susquehanna for over 50 miles, the acid suffocated more than 300,000 fish and affected drinking water and industrial operations, prompting calls for pumps to stop. Although the state fish commission estimated that the river would need three years to recover, not every state agency simply accepted shutting down the pumps, which generated jobs. The Mines and Mineral Industries Department argued that they should restart, otherwise over 1,000 miners might lose work. In December, a month after okaying the resumption of pumping, the state pulled permission because the company failed to comply with its own plan to do the job.[48]

The fish kill sparked a four-year battle to bring acid mine drainage under the provisions of the Pennsylvania Clean Streams Law. In 1965, the state amended the 1937 law to add waste treatment of acid mine drainage, which forced hard coal companies to choose between complying or going out of business. Most major outfits folded, which meant that pumps stopped, which meant that mines flooded.[49]

Market forces doomed efforts to end the anthracite water problem. Hard coal output hit 64 million tons in 1944, but between 1955 and 1961, it dropped from 26 to 18 million. Meanwhile, the industry cut its workforce by more than half, from 33,523 to 15,500, deep mines closed, and stripping operations multiplied. By the time of the AMWCP, economic conditions kept many collieries from fully participating. At the end of 1961, few deep mines operated in the Wyoming Valley; in other fields, miners toiled underground only "above the level of water in abandoned mines."[50]

Officials and engineers could only imagine massive dewatering projects if they believed that anthracite mining could continue. The momentum of history and the large footprint of industry infrastructure may have convinced them that it would, but when they put their faith in complex solutions, they ignored evidence that mining had already all but ended. Transformation had taken place; seeing that it had was tough.

4

Wearing Water

Like many people, I lead a double life: middle-class comforts shield me from the dirtiest effects of an industrialized world, but I sense a deeper, nagging awareness that much material wealth comes at too steep a price, one that no one knows how to lower. With more and more carbon now polluting the atmosphere, the price keeps rising. An obvious sign of my comfort, insulating me from outside, is what I wear, a flow of fashions whose making I used to think little about. Learning to see connections among clothes, coal, and water has compounded my ill ease.

In clothes, we wear water. It may not be visible, but it's there. Economists call this virtual, or embodied, water. Although accounting for virtual water saves water in one place, it signals its loss in another, making the use and commodification of it visible as line items. Early in the twentieth century, virtual water flowed from the Lackawanna watershed not only as coal but also as fabric. Cloaked in waste, the river stitched together coal mining, which gnawed below, and textile making, which unfolded above.[1]

If embodied water makes water visible, then part of the accounting must include hidden costs to laborers, themselves mostly water, whose sweat secures profits for others. In their business practices, coal barons and textile makers too often reduced people and water to virtual status, to a bottom-line existence, which may be why it took until after 1911 for them to accept workers' compensation and until 1969 to recognize black lung.[2]

Coal, clothes, and water call attention to separations of surface and subsurface. Deep mining happens below, out of sight, clothes conceal the wearer, and water blurs the line. Looking into a transparent pool, as Frost does in "For Once" and Merwin does in "The Well," one sees above and below, but not exactly. Refracting light, water opens a gap between what is and what one sees, a dubious doubleness that affects how I read clothes, coal, and the world.

In the late eighteenth century, the water-coal nexus migrated quickly from mining to milling. Tinkering with his steam engine, James Watt refined its capacity for constant rotary motion, an improvement that married water and coal to textiles. As sales to mines dropped in the 1780s, his partner, Matthew Boulton, jumped at the chance to sell engines to cotton companies. When steam went to work in the industry in 1785, it pumped water, but soon it powered machinery. The first automatic device, Crompton's mule drew and twisted fibers to create a thread that it wound around a cop, or quill. In the 1780s and 1790s, the linkage of engine and machine drove sales of steam engines, especially to cotton mills, initiating a worldwide pattern that saw textiles industrialize before other businesses, ensuring that the cotton mill would become a central symbol of the Industrial Revolution.[3]

Historians discuss coal mining and textile making as major threads in the story of the Lackawanna Valley, but they usually study them separately. Interconnections, however, run deep. Mine and mill owners dyed their products; silk makers washed fabrics as much as colliers washed coal; and end-of-shift miners cleaned up and changed what they wore. Both industries also had to control water—seepage in mines and humidity in mills—and family relations knitted their workforces. While fathers and sons coped with make water, mothers and daughters dealt with greenhouse conditions, and before establishing a union, striking silk workers accepted help from the United Mine Workers.[4]

Further knotting coal, water, and clothes, Progressive Era reforms promoted cleanliness. At the height of summer in 1898, with mine acid and sewage polluting the Lackawanna River and swimming in Roaring Brook forbidden, 40 breaker boys jumped into a Scranton reservoir. As they washed in the pond, which supplied drinking water to vexed residents at a Keyser Valley breaker, a half dozen men with clubs tried to run them off and steal their clothes. Forewarned, the swimmers escaped, running pell-mell through woods. One August day two years later, streetcar passengers felt the ties firsthand. As their trolley passed a breaker near Wilkes-Barre, a flume of coal waste came undone, spewing dirty water through open windows, spilling culm across seats, and smearing clothing, "especially the light dresses of the ladies." The mess forced the car back "to the shop for a bath."[5]

Cleanliness and class defined calls for public baths and swimming pools. The *Scranton Tribune* pointed to the "murkiness of the work . . . of the laboring population of the city and its suburbs," and the rival *Scranton Times* agreed, noting, "We doubt if there is a city in this country, or any other, whose working people are engaged in industries that soil the person more than Scranton, and where there are so few opportunities for bathing." Paying a "small sum" for soap, a towel, and a private stall of "perfect cleanliness," the *Carbondale Leader* argued, would erase class distinctions, as happened in ancient Greece and Rome, and would "go a long way toward improving the sanitary and moral condition of towns and cities." Unmoved, local leaders did not dedicate a public swimming pool in Scranton until Independence Day 1909.[6]

Among Progressive Era reformers, miners bathing at home raised eyebrows. Early twentieth-century social workers decried that so many washed in living spaces so open to the sight of children. In 1904, the typical mine worker's home, a two-room structure, was not deemed a wholesome environment, especially when the man must daily bathe in a tub while children are "spectators of [his] nakedness" and when one bedroom accommodates all. With "filth" marking recent immigrants, reformers sought to cleanse the Slav body, which was, one observer claimed, "ignorant of the first principles of cleanliness." But then again, the "Anglo-Saxon" body was, it seems, no cleaner, physically or morally: "The majority of the ills to which our people are subject rise from ignorance of the conditions of health in feeding, clothing and housing the body."[7]

Pennsylvania law required a wash and changing space at each colliery. Company houses often had no running water, and wearing work clothes home kept the places dirty. An in-between space—an extension of home and work—a wash house enabled miners to clean up and go home wearing fresh duds. Drawing on the colliery's main boiler, the wash house, a "particularly hard building to keep clean," maintained warm temperatures for miners' comfort and for drying work attire. Each man had access to two lockers: one for work and one for street wear; at other wash houses, miners dried clothes on hooks that they raised to the ceiling. Although the federal government did not recommend it, some colliery owners put swimming pools in wash houses.[8]

Silk

The local water-energy nexus drew textile making to Scranton. Fleeing New York and New Jersey in the 1880s, silk makers set up shop in the Anthracite Region to escape union demands for decent pay and safe working conditions. Operators believed that they had found a "docile and undemanding workforce" in the area's surplus of unskilled, single women. In Scranton in 1900, women constituted 90 percent of silk workers, with city mills employing 1,600 women and 159 men. The local rail network linked the Lackawanna Valley to markets, especially New York, and owners cut energy costs because they located mills atop the coal that powered them.[9]

Opened in 1873 as a silk throwing business, each month the Scranton Silk mill produced 4,000 pounds of silk, mostly organzine and tram, the "warp and woof of silk goods." Upon arrival at the Sauquoit factory in 1891, raw silk took a two-to-five-hour "bath," which meant washing it in olive oil soap because otherwise it would "cut and cause abrasions on the hard polished steel of . . . the machinery." After the silk was wound onto spools, threads from two bobbins were wrapped onto one tube, an act called doubling; as each doubled thread came from the doubling machine, another device twisted it a number of times before it got wound onto a reel for shipment as filling, or woof. The plant also made fabric; after firms in Philadelphia, Paterson, and New York dyed the silk and returned it, Sauquoit workers wove it into "dress goods, umbrella cloth, or material for silk linings." In 1928, Scranton stood behind only Paterson in silk production.[10]

Plentiful water ensured the right conditions inside mills and made large-scale dyeing possible. With silk sensitive to moisture, owners kept factories as best they could at 70–75 percent humidity and a 70°F temperature. To create these hothouses, operators sited them along streams and rivers, installed humidifiers, or pumped steam into rooms. A moist atmosphere kept mills running smoothly because dryness snapped threads and made silk "susceptible to electrical influences." Too much humidity also caused trouble, however, so hot and humid days might require building "steam in the heating coils so as to dry things out." One observer of the industry dismissed concerns that additional heat might hurt workers by pointing out that drier air meant evaporation of more sweat, which "tends" to cool bodies. I wonder whether he had ever worked in these conditions for a typical week of 10-hour days.[11]

A water-coal-silk connection, dyes were derived from coal tar, a byproduct of coal gas, itself usually a byproduct of burning bituminous. Developed in 1856, artificial dyes prompted textile makers to color fabrics in vivid hues and publishers to add tints to pictures in books. By 1918, anthracene, a hydrocarbon in coal tar, also made dyes, but problems with its purification limited its use.[12]

Dyeing silk needed "astonishingly large" amounts of water, the purer, the better. An outfit coloring 20,000 pounds of silk per day required seven million gallons. Dye houses also used "large volumes of soft water to feed their boilers," and owners sought to fire steam engines with anthracite because it burned to little soot.[13]

Most silk operations in Lackawanna County were throwing mills. Standing long hours at loud machines, throwers were mainly young and female. In Carbondale at the turn of the twentieth century, the Klots Throwing Company mill on Belmont Street regularly employed 450, 60 percent of whom were female; only 5 percent were over 21, a handful were children 11 or younger, and two-thirds were 15 to 17 years old. In all of Carbondale in 1900, the average and median age of silk workers was 16. In 1907, the Klots mill ran a "six-day, fifty-seven-hour work week (ten hours during the week and seven hours on Saturdays)."[14]

In 1900–1901, long hours and low wages led the Klots workforce to strike, a labor action that spread to mills elsewhere, including Sauquoit. The local United Mine Workers pledged support for the "claims of the children," who explained that special machinery at the mill ran at a rate that demanded "double the attention"; standing for "twelve hours a night and under the glare of electric lights [was] too much"; beginners worked "from two weeks to six months without any remuneration"; and frequent fines ate earnings that adults knew were insufficient for self-support. In February, an arbitrated wage increase settled the Klots strike, which had lasted eight weeks. After the much-larger strike ended in April, the superintendent of the Sauquoit mill pointed out that silk businesses would likely not expand in Scranton; new mills would be built in the South. Events proved him wrong in the near term but right in the long run.[15]

Without really realizing it, I touched textile making at two workplaces: farm and factory. Although the farm depended on dairy cows, we raised a handful of sheep, about 20. They grazed in the Sheep Pasture, land a mile and more from the farm. My father told of times when he and his father brought the merinos further, to Sheep Falls at Niagara for washing. As a child, I once helped bring the flock to pasture. My father led them, and I followed, at a sheep's pace, up the dirt road and across Route 371. After dogs took too many, the ewes stayed on the farm, where they wandered the ground behind the garage and in the orchard behind the house. In winter, they crowded the Sheep Barn, sharing space with calves and heifers. In March, we'd sometimes have newborns in cardboard boxes beside the kitchen radiator. The last wool went to North Carolina.

In an early off-the-farm job, I worked part-time at the Klots mill, by then called Gentex, which was an education in learning that I was not cut out for factory work. A major employer in Carbondale, and just across the Lackawanna River from Doyle & Roth, Gentex relied on military contracts. When the U.S.

waged war, the plant usually ran full tilt. If not, it sometimes didn't. Working two weekend shifts, I helped to make army helmets. I lasted two months, October and November, entering and leaving the plant in darkness.

For 12 hours, I sat at a machine that cut Kevlar. Spitting out 72 pieces per minute, the machine breathed in sync with my pulse. I caught pieces in my palm, set them aside; caught pieces in my palm, set them aside. I couldn't help but watch the clock, which stared from the far wall opposite me. For someone used to moving freely, sitting in the same place was agony. I experienced what Crompton's mule introduced to the world: work at a machine's pace.

Windowless, the room was dry, very dry. Three breaks marked the day: 10 minutes in the morning, a half hour at noon, and another for dinner. Loading a roll of Kevlar into the cutter constituted the only variation in the routine. Rarely would the machine break down, which when it did was a relief. Around the room, others sat at similar devices doing similar work. Although I earned a little more than minimum wage, I soon added to the place's turnover. I can't remember whether I quit or was let go, but I don't think that I was fired. I could leave, but not everyone could or would.

Just as a clean appearance mattered in marketing clothes, so it did in selling coal. Colliers showered anthracite with water as it left breakers to make it "clean and shiny." Spraying created a "metallic look" that "help[ed] convince buyers that hard coal was superior to soft coal." But not all users appreciated the look. Some distrusted shiny anthracite as "*too clean*" because, they believed, "coal containing more pieces of dull appearance . . . burn[ed] longer." No matter how much it was sprayed, though, coal remained dirty, which steered some customers to oil and gas.[16]

When anthracite producers discovered, in the 1920s, that spraying was not enough to sell coal, they added dye to the shower. The Glen Alden Coal Company colored coal blue to distinguish it from competitors, Reading Anthracite concocted a red dye, and Hudson Coal sprayed its silver, advertising it as "Sterling Coal," the best. The Pennsylvania Coal Company and Pittston Coal marketed "Jet Black Anthracite," and another operator painted its product pink. Although a marketing tool, dyeing may have started as a way to keep the poor from stealing anthracite from coal cars.[17]

Staining the Lackawanna, dye flowing from breakers mixed with colors loosed from textile makers. In the 1950–1960s, a carpet factory next door to North Scranton Junior High School every so often released leftover dye that painted the river "red, green, blue . . . silver, pink, grey, beige." During the same decades and several miles upstream, a similar palette played out: "The river in those days was all different colors, green, black, brown, grey. . . . Never was there any clear water."[18]

Scranton Lace

Lace curtains appeal to a buyer's desire to signal progress in fulfilling the American dream of material success, a wish captured in the often-pejorative term "lace-curtain," meaning "aspiring to middle-class standing." A display of wealth, a window curtain, like clothing, marks and blurs a line between inside and outside, shielding one from the world. Making private a wash place, a shower curtain not only draws a line between wet and dry but also hides the unclothed, luxuries unavailable in many early mine worker homes.[19]

Unlike silk throwing mills that spun thread, Scranton Lace wove it. Looms meshed three kinds: warp, spool, and bobbin yarns. To discover imperfections in curtains, women in a "'reading' room" examined each "as a proofreader would a column of news matter." The cotton curtains then underwent bleaching, starching, and several washings before ironing, building, and pressing. The lace-making machinery used powdered graphite, which "enabled parts to move against each other with silky smoothness—but blackening everything, machine, workers, and lace, in the process. Only after the lacework was completed was it treated to a 'bath' restoring it to a white color."[20]

Scranton Lace adapted to economic instability as best it could. Easy access to coal and labor brought lace production to Scranton, and at one time, the company owned cotton mills and coal mines. Employing about 275 in 1897, the plant produced 1,000 curtains per day on 15 looms. Early in the twentieth century, its 1,400 employees led the nation in making Nottingham lace. During the Depression, the owners turned to tablecloths, and after World War II, they converted a unit from the wartime production of parachutes to the assembly of vinyl shower curtains. For openwork designs, the factory completed all processes inhouse: "designing, bleaching, dyeing, starching, mending, hemming and packing." I've heard that competition from China killed the final order, in 2002.[21]

In 1927, mining threatened the laceworks. Built beside the Lackawanna River, the factory sat atop 22 acres of coal. Of the seven veins below, only the upper one was solid, the second was half removed, the third two-thirds taken. In the latter, crews had almost finished backfilling, and the company planned to pack the second tunnel within five years. Water filled four lower veins, "render[ing] them not only inaccessible but very little danger of any cave trouble." Once the second vein was refilled, the company expected "no surface trouble." The business would finally stand on solid ground, despite its watery depths.[22]

Phoebe Snow

The Progressive Era focus on cleanliness affected how companies marketed anthracite. In the days before air conditioning, railroads that burned bituminous coal dirtied passengers because soot and smoke invaded cars through open windows. Burning hard coal, however, created little soot, smoke, or cinders. In the early 1900s, the Lackawanna Railroad marketed itself using the "theme of cleanliness," which was "at least partially influenced by a message Mark Twain had written to the railroad in 1899 after travelling to Elmira when he wrote, 'Left New York on Lackawanna Railroad this A.M. in white duck suit, and it's white yet.'" Clean clothes sold coal.[23]

To advertise anthracite as clean, in 1901, the railroad invented Phoebe Snow, whose name associates her with virgin waters and whose immaculate white dresses stayed pure as she traveled on trains that hard coal powered. The company placed the first ad on New York City trolleys. Targeting middle-class anxieties and aspirations, it featured a woman in white beside a Lackawanna passenger car and included a poem: "This is the Maiden all in Lawn / Who boarded the train one early morn / That runs on the Road of Anthracite / and when she left the train that night / she found to her surprised delight / hard Coal had kept her dress still bright." "Maiden all in Lawn" limited rhyming, so in 1902, the company married her aboard a Lackawanna train to Mr. Snow. Promoting travel and tourism, Phoebe visited towns and vacation sites along the rail line, playing a large role in the growth of the New Jersey suburbs. In the first decade of the ad campaign, the railroad saw its passenger numbers jump 80 percent. Copywriters never tired of rhyming "white" with "anthracite," which relied on the binaries white/clean and black/dirty.[24]

On her travels, Phoebe Snow sold clean drinking water. When the conical paper cup came into vogue, she advocated the sanitary nature of the throwaway container. Ads in 1909 showed her sipping from one: "On railroad trips / No other lips / Have touched the cup / That Phoebe sips. / Each cup of white / Makes drinking quite / A treat on the Road / Of Anthracite." Underscoring "comfortable and luxurious accommodation" on the railroad, an ad in the *New York Times* proclaimed, "Individual Drinking Cups as a part of its regular service on all <u>through</u> trains. This innovation means that you can obtain a cool drink of water whenever you want it and under perfect sanitary conditions."[25]

Phoebe Snow also endorsed burning anthracite for a cleaner environment. Digging at a hard coal rival, a refrain in ads made anthracite the protector not only of clothes but also of land: "it means that there is absolute freedom from black, choking soft coal smoke that soils the clothes and mars the landscape." When one need not worry about taking "a bath immediately after leaving the train," one can "keep one's eyes on the passing scene . . . for every mile is picturesque." Jingles reinforced this message: "Each passing look / At nook or brook / Unfolds a flying picture book, / Of landscape bright, / Or mountain height, / Beside the road of anthracite."[26]

Although the ad campaign ended in 1917, Phoebe Snow returned every so often. She made an appearance in 1930 in "traditional white to represent the cleanliness of the new electric trains," and during World War II, she donned an "olive drab uniform to further the efforts of the War Department." In the 1960s, the Erie-Lackawanna Railroad resurrected her to entice people aboard its trains, noting in promotions that in her early years, she "became widely known as the symbol of clean and gracious travel on the Lackawanna," "whose locomotives took the 'sin' out of cinders by burning hard instead of soft coal."[27]

A public face of the railroad, Phoebe Snow cloaked darker elements of coal and clothing. Portrayed as clean, middle class, and female, she contrasted sharply with the stereotypical image of the dirty, working-class male workers who mined the coal that sparked her birth. Her spotless dress also separated her from the stained and patched clothes of the working-class women and girls who worked at throwing mills and lace factories. Although a nod to the New Woman and sporting a middlebrow fashion taste, she maintained an aristocratic air associated with elite business travel.

To wear means to bear or to carry, as we do with clothing, but it also means to deteriorate through use. Wearing water leads to water wearing us. Although the virtual water in Phoebe Snow's dress may have helped to present her and anthracite as clean, clothing is as dirty as coal. To make 70 million tons of fiber in 2007, the global textile industry burned 145 million tons of coal and absorbed 1.5–2 trillion gallons of water. These days, clothes make up 10 percent of world carbon emissions, comprise the world's second-largest polluter of fresh water, and soak up 25 percent of manufactured chemicals. In a toxic loop, plastic fleece makes its way into seas, accumulates in marine life, and returns to us through the food chain. A metaphor for appearances, clothing lays bare the artifice in arguments for clean coal and natural gas: garments manufactured with fossil fuels stay dirty, no matter how one greenwashes them.[28]

5

Fouling Water

The dirty water that drenched trolley passengers in 1900 streamed from the Prospect breaker in Pittston. Culm had clogged the operation's flushing system, pushing wash water into the sluice that was to carry the waste to the Susquehanna River. By the time of the accident, overflow had piled culm about 200 yards out into the channel, nearly half its width, and almost 300 yards downstream, threatening to kill fish, block sewers, and increase flooding, not to mention deprive people of relief from the August heat. Responding to complaints about damage to the river, a company official took a common corporate stance: "the culm does not pollute the water but what little of it that has gone into the river from the colliery has benefitted it."[1]

During the first decades of the twentieth century, the anthracite industry projected a public face of purity—clean coal and spotless train travel—that masked how mining dirtied water and scarred landscapes. Boosters promoted hard coal as clean in two ways: (1) smokeless because it burned to almost all carbon and (2) deliverable free of dirt and rock. Rather than concede that coal polluted waterways, colliery owners and local officials argued into the mid-twentieth century that mine acid neutralized the effects of sewage in local streams and rivers. At the same time, water-treatment facilities used anthracite as a filter, and promoters today tout its high carbon content—its purity—as an environmental advantage because of the coal's high heat output relative to other fuels. Defining "clean," though, has never been straightforward, especially when applied to fossil fuels. What I burn reminds me that I'm not as clean as I sometimes think I am.

Scranton pushed the clean-coal narrative. At the turn of the twentieth century, the city board of trade argued that the U.S. Navy should power ships with anthracite because billows from burning soft coal revealed a vessel's position and obscured signal flags. Aboard warships, they asserted, smokeless coal ought to be as necessary as smokeless powder. In 1907, Admiral Robley Evans publicly endorsed the idea, urging the federal government to seize anthracite for use mainly in the navy, despite the steep price: $18 billion. In taking the tradesmen's argument another step, Evans ignited blowback for his "fantastic scheme."[2]

The Scranton Board of Trade wanted the U.S. government to buy anthracite, not to seize it. Evans commanded the North Atlantic Fleet, so coal and rail companies feared that his idea might take concrete form. Challenging the admiral, the Lackawanna Railroad enlisted Phoebe Snow to preserve anthracite for private burning. The company derided the mariner's suggestion in a *New York Tribune* ad. Towering at the end of railroad tracks, Phoebe defies a dwarf Evans, who stands aboard a tiny battleship. Between them runs a shoreline that echoes the East Coast; coal banks substitute for the Appalachians. Phoebe threatens him, "So you would take my heart's delight! / You little know / The friends I have, / Who ride the Road of Anthracite." One passenger likely interested in killing the proposal was financier J. P. Morgan, whose banks backed the railroads that ran the mines.[3]

Washing Coal

To dress coal, collieries first separated it from whatever was not coal, mainly rock, in a process called washing. Cleaning began inside mines when workers separated anthracite from slate, often tossing the rock into worked-out rooms. Breakers, developed in the 1840s, used spiked rollers to crush coal, perforated screens to size pieces, and a dry method of washing: boys picked rock from chutes as the coal slid toward railcars. In 1904, teams of 10-to-14-year-olds still removed slate from anthracite, despite child labor laws and improved methods of washing.[4]

Machine cleaning with water likely began in the 1870s in the southern field, and by 1895, collieries in the Wyoming Valley had turned to wet methods, which took three to six tons of water to wash one ton of coal. In the 1920s, many operators wet washed using a sand flotation separator, the Chance cone. As coal passed across shaking screens, water jets separated large and small sizes. Coal is lighter than rock, so the cone contained a "uniform mixture of water and sand . . . with a specific gravity slightly greater than that of anthracite and less than that of the impurities." When coal entered the agitating cone, rock moved to the narrow bottom and was drawn off; coal floated on top and was skimmed away to sizing screens. To catch leftover anthracite, the rock was washed again. After sizing, coal passed through "cascades of water" that removed silt and sand because anthracite sold "almost entirely on its appearance." Many collieries dewatered coal, clarified the water, and recirculated it for reuse because a constant supply of fresh water would have been too costly. In the 1940s, though, one anthracite company "purchased 20,000 acres of watershed . . . to guarantee a supply of fresh water for washing coal."[5]

Companies also used jig preparation and the Rheolaveur system. In the jig process, coal entered a partially divided, water-filled box. On one side, coal spilled across several inclined, perforated plates. On the other, a plunger moved up and down, pushing and pulling water back and forth through the holes. As this happened, rock slid to the base of the incline and was removed through a gate; coal rose to the top and was carried off in a flow of water. In the less used Rheolaveur system, water conveyed coal through a series of troughs. The flow brought coal to the top and rock to the bottom. At intervals, pockets in the trough carried away the rock; water jetting up from each pocket kept the coal from following.[6]

Unsellable coal dirtied waterways in at least two ways. Wet cleaning produced wastewater, which collieries directed to rivers and streams. Although companies sometimes mixed rock and small-sized coal with water and flushed it underground, most often they simply piled the waste, which rain carried off. Chunks of time broken and piled, culm banks polluted waterways, grew into eyesores, and poisoned not only the Lackawanna River but also the Susquehanna and the Schuylkill, which yearly absorbed almost a million tons of waste coal, with culm banks contributing 30 percent of the total. In 1945, the U.S. Army Corps of Engineers estimated that the Schuylkill contained 26 million tons of silt.[7]

Visible reminders of coal washing, culm banks have come and gone and returned. The earliest piles held a lot of recoverable coal. To reclaim it, operators in the northern anthracite field built culm-cleaning systems called washeries. Although washeries removed many banks 100 years ago, their mining still left waste, which itself was piled. Today, companies level these heaps to build on or truck them away to fill gaps elsewhere. This ongoing effort to rework worked-out land depends on a federal tax on coal operators; in the Eighth Congressional District in northeastern Pennsylvania, "more than 300 abandoned mine land sites . . . require rehabilitation that will cost at least $114 million." Climate change endangers future funding: less reliance on coal will mean less tax revenue to support the remediation of mine-scarred land.[8]

Bodies of Water

On a typical day during the school year, I walk 10 minutes to my office, teach classes, attend meetings, and walk home. On this loop, I hardly ever touch ground, only concrete and macadam, marble and steel, carpeting and tile. My daily round reveals not only my separation from nature but also my total absorption within it. The disconnection is superficial—my feet don't come in contact with soil or grass—but the enmeshment is deep. I breathe and exhale, my sweat evaporates, my muscles burn—but it goes even deeper: I carry within me millions of microbes. I may want nature beyond my skin, but my body tells me differently—and not just that nature is under my skin. I am nature; I am water. Largely a carbon-water nexus, my body is about 60 percent water. My thinking swims in it, and without it, I am a waste pile, of about 67 percent carbon. A flexible grid of wet bone, muscle, and skin, I walk to campus along ruled streets that run atop sinuous grids of water-filled mines.[9]

Scholars who question the divide between the human and the nonhuman point to the human body as a place where the two are one. The modern body, enclosed and isolated from the environment, has given ground to the ecological body, which is coextensive with the environment. Whereas the modern body seals itself against the water sack that is the earth, the ecological body remembers that in the womb it floated within a small world of water. Transcorporality asks me to pay attention to my body as embedded in its surroundings. An everyday example: drinking water.[10]

A reservoir of water, a coal miner knows firsthand workplace interpenetrations of body and environment. An explosion can break eardrums, crack bones, and embed coal in skin, and underground labor blackens bodies not only outside but also inside. Wash houses may help miners arrive home "as clean and well-dressed as workmen in any other industry," but their clean appearances often mask lungs caked with coal.[11]

As easily as bad water can kill a river, it can kill people. At the end of the nineteenth century, conventional wisdom claimed that running water washed itself, and rivers do self-purify in three ways: oxidation, deposition, and dilution. If a stream has sufficient dissolved oxygen, bacteria break down organic matter enough that the stream might run clean. In deposition, pollution settles to the river bottom, and a large enough volume of water can dilute toxins. But waterways can be overwhelmed.[12]

Cities that relied on rivers to wash away sewage—out of sight, out of mind—suffered typhoid outbreaks. In 1890, deaths in Lawrence (83) and Lowell (150), Massachusetts, tallied close to rates in the 1832 London cholera epidemic.

Opening a water filtration plant in 1893, Lawrence cut cases by 80 percent. This success proved that waterborne bacteria spread typhoid, undercutting arguments that rivers always self-cleaned: flowing water was not as safe as assumed.[13]

Typhoid also plagued the northern anthracite field. In 1885, Plymouth, a town of 8,000 near Wilkes-Barre, endured an epidemic that sickened 1,200–1,300. People in one area blamed typhoid on drinking water contaminated with culm, and in 1905, the illness swept through nearby Nanticoke, killing at least 4 and sickening 132. In December 1906–January 1907, Scranton suffered an outbreak that affected 1,171 residents and killed 123. Three years later, in August 1909, it struck again, landing at least 28 people in four city hospitals.[14]

Mine Water: Sewage Solution

Infrastructure sells. When land developers Jordan, Hannah, and Jordan marketed lots in Richmont Park in the early twentieth century, they highlighted not only our Scranton neighborhood's graded street, bluestone sidewalks, and convenience to streetcars but also its gas, water, and sewer lines. Advertisers promoted the "perfect sewer system" that kept the place "clean and beautiful," a pipe that reached the river.[15]

For decades, people simply sent waste from mines and homes "away." Not only did Scranton use the Lackawanna to rid itself of sewage, but local collieries and textile mills relied on it to wash away acid and dyes, a strategy that passed poisons downstream. A similar attitude held sway elsewhere, but here it had a twist: coal operators argued that mine water neutralized sewage. Just as "away" is not a strategy for dealing with pollution, neither is arguing that one river-killing problem solves another.[16]

By the early 1890s, the "away" philosophy wasn't working well. The town health officer told the city councils in 1889 that "in spite of our bad streets and lack of sewers, the health of the city has been remarkably good." Acknowledging the Lackawanna River as "our main sewer," he then turned to the "garbage question," wondering whether "even moderately dry times" would make conditions "terrible." They did. In 1891, city officials approved Stafford Meadow Brook as an open sewer from Pittston Avenue to the Lackawanna, allowing "breweries and other industrial establishments" to dump at will. A year later, a crew at work along the brook's banks pointed to "filth" a foot deep or more in the streambed. Sun and still water made the "smell . . . almost unendurable."[17]

Scranton made it illegal for scavengers to dump waste into the river, but municipal leaders refused to fund alternative ways of disposal. In 1892, ordinances required cesspools to be cleaned up. Licensed scavengers answered the call, but the city designated no place for them to unload what they collected, and the city councils balked at building an incinerator. Scavengers had two choices: keep their carts loaded or dump the contents illegally.

Wagonloads went into the Lackawanna. The "night soil of the entire city" slipped into the river at Park Place; residents blamed the dumping for four deaths. The city health officer called the water an "abomination . . . [with] in some places a foot of fertile matter in the bed of the stream." The situation grew so bad that only a "marvelous climate and the immense amount of anthracite that is burned" saved the city from a "terrible plague." Although Scranton barred residents from dumping feces into the river, the city passively encouraged coal companies to pour acid into it. State lawmakers felt more strongly. When they passed the Purity of Waters Act (1905) to forbid new sewage systems in Pennsylvania from dumping waste into streams, they excluded mine water.[18]

"The River": An Adult's View

Teaching me to look with both eyes, a poem is a place where separate spots of time can unfold at once. Paul Kelley's "The River" invites readers to compare adult and child views of the Lackawanna. This double vision encourages readers to wonder how best to see the river.[19]

The adult understands the Lackawanna through maps and books. With a wider perspective than he had as a child, he knows its course, its connection to other waters, and its change from "Where waters rushed" (1) to "a slow moving stream" (37). Nodding to this bird's-eye view, the poem represents the river's flow in its sinuous shape, which lists towns, named mainly for men in local rail and coal industries. In "drawing these towns together like knots" (22), the river crosses corporate divides, connecting places in a shared history whose foul effects it carries "into / the Susquehanna . . . & on, to the Chesapeake Bay / & the sea" (24–29). This map-like movement of "sewage, & refuse, & waste" (38) also charts a problem only the adult sees: the towns pollute the Atlantic.

The adult keeps his distance, knowing the Lackawanna through his reading. Nodding to the Native American "*Lechau-hanneck*" (5) and "the Shawnee & the Moosic / Mountains" (30–31), he thinks in clichés borrowed from written descriptions: "'deer and moose' / 'among the laurel, and the hemlock'/ 'and the pine'" (32–34). The empty phrases confirm for him the loss of what he never knew, "the river / abounding once with fish" (35–36). Unable to bear the thought of the real river, which "now offends" him (39), he turns to memory, recalling how he and his childhood friends enjoyed the river as a "joke: / the 'Lacky'" (40–41), a place of "stink & shit" (52), a more exact description than the adult's euphemistic "sewage, & refuse, & waste" (38). He and his buddies played with what they found. They got dirty. If the adult were to do the same, knowing what he knows now, he might imagine himself a river keeper, not its mourner.

Coal operators and Scranton officials claimed that mine acid would solve the sewage crisis, but this faith only allowed both afflictions to fester, unchecked, despite complaints and dead streams. When the city health officer declared in 1897 that "sulphuric matter in the water renders [sewage] innocuous," others responded that "the olfactory organ will furnish evidence apparently to [the] contrary." Despite their senses, some may have accepted the coal-cancels-sewage assertion because it kept coal mining clean, a faith that required bracketing the Lackawanna's smell, appearance, and lifelessness. In 1902, it was still for some an open question "whether a stream polluted by sewage is assisted or retarded in its self-purification by the sulphur-polluted tributaries." Observers suggested that mine waste "may have an important chemical influence upon sedimentation, if not an appreciable effect upon the oxidation of organic matter." Delaying action, the debates maintained the status quo.[20]

Science not only supported the argument that mine waste neutralized sewage but also revealed a related problem. In 1904, the U.S. Department of the Interior found more organic matter in the Susquehanna at the upper end of the Wyoming Valley than in the lower and attributed the difference to "large amounts of mine waste which are turned into the stream." But when fine coal fused with sewage and fell to the river bottom, it raised the bed in some places as much as 12 feet, a "contributory cause of recent damaging floods in the city of Wilkes-Barre." Unfit for domestic use even without the waste it carried, the river might as well continue as a sewer, the report noted, rather than be purified, despite the flooding risk. Happening unseen, a rising river bottom was another out-of-sight/out-of-mind problem.[21]

Attention to connections between coal and sewage did not disappear with the end of deep mining. Noting in 1957 the "gross contamination of rivers," Conowingo tunnel planners suggested that the decline of hard coal could clean up the sewage problem. They recommended abandoned mines as disposal systems: "there is no good reason why the present discharge of sewage effluent into the vast worked-out areas and pools in abandoned mines could not be greatly increased, thereby removing much of that pollutant." One way to clean up a surface river, this argues, is to pollute a subsurface one.[22]

Former Scranton residents remember the coal-cancels-sewage argument. In the 1940s, people accepted that mine acid made sewage harmless. Pointing to the burden on taxpayers, local government officials even promoted the idea to dismiss demands for a sewage treatment plant. In 1944, city officials offered the acid-kills-bacteria argument to challenge state enforcement of anti-pollution laws; the real reason for their opposition, however, was because sewage plants were a "tremendous cost" and keeping the Lackawanna acid free "would bankrupt the anthracite industry." In the 1950s and 1960s, this thinking kept the river "a stinking, putrid mess which was made worse, 24/7, by a large minimally treated sewerage outflow just north of the Albright Ave. Bridge," very near Scranton Lace. The sewage killed fish, frightened off birds, and fed vegetation along the banks to grow a "river-garden . . . one enjoyed at a safe distance." Others offered the coal-sewage argument in reverse. Responding to state efforts in the 1940s to stop the discharge of untreated sewage into the Susquehanna River, a Wilkes-Barre city engineer claimed that sewage "served a beneficial purpose—it counteracted the acid in the river."[23]

In 1962, the year that a crew drilled the Old Forge borehole, a Pennsylvania Health Department engineer reported to the Scranton Sanitary Water Board that 90 percent of homes in the city were dumping about 10 million gallons of sewage a day into the Lackawanna River. Following up on an idea floated by a city councilman, a city solicitor informed the council in 1963 that any "suggestion of disposition of sewage by boreholes into the mine" would be illegal. Although the city formed a sewer authority in 1966, Scranton led the list of sewage polluters of Chesapeake Bay in 1991. In 1999, state officials discovered that 145 homes in neighboring Taylor were discharging raw sewage through a borehole directly into coal tunnels.[24]

By 1973, wastewater treatment systems that combined sewage and storm water operated throughout the Lackawanna Valley. Heavy rains, however, sometimes overwhelmed the facilities, which responded by letting sewage into the river, which overpowered modern sewage treatment's "Magic mile," the "distance it is commonly believed to take for a river to disperse remaining pollutants sufficiently, or absorb them via aquatic plants and microorganisms." In 2012, the city sent 700 million gallons of combined sewage and storm water overflow into the Lackawanna and on to Chesapeake Bay. To protect the bay, the U.S. Environmental Protection Agency (EPA) sued the city in 2009 to force it to separate sewage and storm water; a 2013 settlement required $146 million in system upgrades over the next 20 years. So far: mixed results.[25]

"The River": A Child's View

While the adult's view of the Lackawanna River is distant and abstract, the child sees it up close, as a particular place. He details his lived experiences of it, telling us that he and his friends roamed "at the foot of Court St. / near

Ella Burns's bar / under the bridge at / Albright Ave." (48–51), in the shadow of Scranton Lace. Here they play with the river's "rusted out junked / cars" (44–45), which they drive, imaginatively making the most of what adults have worn out and tossed aside.

The children add loss to their play. Aware that "a baseball / homered into the water / was unusable" (52–54), they decide that the fouled ball is "an automatic out" (56). A lost ball may not end their game, but it does change how they play. A batter might power back on his swing or shift his stance, depending on home plate's distance and direction from the water. An out, the wet ball symbolizes a penalty for violating a limit that the kids devised to accommodate the polluted river. Played within these limits, the game can go on indefinitely. Similarly, if adults worked within limits of the natural world, we might long sustain ourselves. If nothing else, we might put off losing. Nature bats last.

"The River" reveals a double innocence when read in the context of climate change. Published in 1981, before major efforts to clean up the Lackawanna, the poem calls attention to poisoning of the river, and one could read the adult as judging his childhood play near it as naive and dangerous. Read today, however, with the river much healthier, the grown-up looks naive, endangering himself in fretting only about what's lost and doing nothing to help. An innocent, the child adapts to wrongs adults have dealt him, unaware that polluting the river—or the atmosphere—might be mitigated, if not ended. As an adult, he knows better . . . or should.

Clean Coal

Between World War I and 1930, hard coal sales dropped 30 percent due mainly to competition from fuels other than oil and natural gas. Anthracite especially felt heat from coke and manufactured gas, fuels that also cloaked themselves in cleanliness. All three industries paraded workers in white to mask their dirt, and in 1914, the manufactured gas industry introduced Nancy Gay, a rival of Phoebe Snow. By 1917, gas competed with anthracite on price, but coal companies responded slowly. Despite efforts in the 1920s to recapture markets through "improved testing for size and purity . . . new flotation methods and . . . major investments in breakers and washeries," hard coal lost the cleanliness battle to gas.[26]

The meaning of "clean" matters in energy policy. In a 2007 ruling based on the Clean Air Act, the U.S. Supreme Court defined carbon dioxide as a pollutant; two years later, the EPA declared CO_2 a public health danger. Defining carbon dioxide as a pollutant gave the EPA the power to control its production, a fact fossil fuel companies bristled at. In 2015, the EPA finalized the Clean Power Plan, which the agency based on the Clean Air Act. Designed to move the nation toward increased reliance on cleaner energy, the plan hit a snag in 2016, when the Supreme Court stayed implementation, pending review. In 2017, President Trump signed an executive order meant to gut the plan, undo mining regulations, and open federal lands to more mining. Ignoring climate change, the order in effect authorized loading more carbon into the atmosphere. In 2020, the Trump administration undid methane regulations, an announcement made in Pittsburgh, in a state whose constitution guarantees every resident a right to clean air and clean water. Although opposed by major oil and gas companies, the rollback would have dumped an additional 400 million tons of carbon dioxide into the atmosphere annually. The next year, Joe Biden signed a bill that put the methane regulations back in force. Reversing a policy reversal reveals a nation unsure of its response to climate change, which presses on.[27]

The definition of clean energy is fuzzy enough to create confusion. Some tout nuclear power as clean, despite the difficulty of disposing of its waste. Natural gas is clean, companies claim, or at least it's cleaner than coal and oil. Although gas contributes less to climate change than does coal, the amount is not negligible. It's as if the nation needs alchemists to transform fossil fuels into clean energy so that we can comfort ourselves that climate change is not our doing.[28]

Greenwashing, or pretending a product or service is environmentally friendly, hides pollution and destructive practices; it makes something appear to benefit the environment when it does the opposite. Coal companies in West Virginia argue that mountaintop removal, ripping away the tops of mountains to access coal, allows them to reach "lower-sulfur—and thus 'cleaner'—coal," making the practice of destroying mountains into "green" mining, an argument that many legislators and voters accept. Between 2007 and 2010, few people knew that Chesapeake Energy, a major natural gas company that operated in northeastern Pennsylvania, and the Sierra Club, an influential environmental organization, promoted natural gas through cofounding the American Clean Skies Foundation.[29]

Companies greenwash anthracite. Blaschak Coal Corporation, which operates the Lattimer breaker near Hazleton, promotes anthracite as clean burning. A local seller of coal-burning stoves describes hard coal as "one of the most clean-burning hydrocarbon fuels out there. For the amount of heat you get in a residential home, there's really a very small carbon footprint." Sidestepping that burning anthracite loads carbon into the atmosphere, the word "clean" may suggest no harm to the environment, but it does, in fact, mean that almost all the carbon goes up in smoke, leaving little ash but contributing to global warming.[30]

When people talk about clean coal, I hear greenwashing. It may be possible to collect CO_2 "produced at coal-fired power plants," but you're still left with CO_2, which is the problem. Storing it underground only postpones the day that it moves to the heavens. Coal is not clean but dirty in several ways: mining piles up environmental, social, and economic costs. To mask this, in 2010, the coal industry spent $40 million to promote coal as clean energy. In his 2018 State of the Union address, President Trump, a self-proclaimed fossil fuel fan, asserted that in his first year in office, his administration had "ended the war on beautiful, clean coal." In 2023, the U.S. Department of Energy pushed ahead with carbon capture demonstration projects, despite spending over $1 billion since 2009 with limited success.[31]

In a weird loop, anthracite purifies water that mining pollutes. Although mainly burned for home heating and steel making, anthracite also aids water filtration systems. At least as early as 1918, treatment plants used anthracite, which "worked as efficiently as sand." Carbon Sales, in Wilkes-Barre Township, today specializes in Anthrafilt, a water filter made from anthracite mined in Tamaqua, where the "coal averages from 85 to 91 percent [carbon] which means it's nearly pure." In the 1950s, researchers at the Hanford Nuclear Facility in Washington State found that "filters made of anthracite and sand together [were] superior to filters made of either material alone."[32]

Distributing clean water is a world problem. Over 1 billion people have no sure access to it; over 2.5 billion have no basic sanitation. A result: waterborne illnesses kill 6,000 children per day. Ballooning urban populations increase pressures on water availability, and expanding access to drinking water will lead to a 30 percent jump in demand for fresh water worldwide by 2030.[33]

In the United States, sewage, chemicals, and lead contaminate drinking water for millions. About 800 cities with a total population of 40 million use combined systems that discharge about 850 billion gallons of untreated sewage and storm water into "inland and coastal waters." The chemicals polyfluoroalkyl and perfluoroalkyl (PFAS), found in many consumer goods, also affect communities across the nation, poisoning the water of more than 110 million. As happened in Flint, Michigan, pipes in many systems leach lead into drinking water, causing "one of the great underappreciated public health disasters of the modern era."[34]

Responding to a long history of dirty water, the United States passed the Clean Water Act in 1972. The law was important for the Anthracite Region because as the coal industry waned, underground mines became a "dumping ground for industrial wastes," ensuring that the region's colonial status didn't end when mining ended, it merely changed form. In 1979, investigators traced an oil plume in the Susquehanna to the Butler Tunnel, which emptied mine water into the river at Pittston. The waste slick extended at least 60 miles and may have reached Harrisburg. The interstate highway system that intersects near Pittston enabled tankers to have easy access to a truck stop that boasted a borehole originally drilled to direct raw sewage into mines. For several months, millions of gallons of used oil, cyanide, and other chemicals went into underground workings that the tunnel drained. The cleanup was so expensive that the disaster became "the poster child" for the creation of the Superfund program. Although the EPA declared the initial mop-up a success, in 1985, Hurricane Gloria flushed out more of the toxic mess, poisons that originated in cities that had abandoned anthracite to burn its fossil fuel cousins.[35]

The day before a judge sentenced the people responsible for the Butler Tunnel disaster, a grand jury charged a Scranton man with dumping more than 3.5 million gallons of toxic wastes into mines under the city. Between 1976 and 1979, the accused allegedly poured about 581 tanker trucks of the waste into a floor drain inside a Keyser Avenue garage. The drain connected to a borehole that dropped 110 feet into an abandoned mine. The man collected some of the wastes from companies in Philadelphia, the first major market for anthracite.[36]

As important as the Clean Water Act is, it's not set in stone. In 2020, the Trump administration weakened the act and its counterpart, the Safe Drinking Water Act (1974), when it lifted protections for ephemeral waterways, streams that run only after rain or snowmelt. Drought-stricken regions in Western states rely on these waters for drinking: ephemeral streams that feed the Rio Grande help supply water to millions. The revision affects nearly one-third of the nation's drinking water and undoes protections for "several million miles of streams that feed larger bodies of water and about half of the nation's wetlands that serve as natural flood protection and natural filters for drinking water." At the same time, Trump also loosened restrictions on how "coal-fired power plants dispose of wastewater laced with dangerous pollutants like lead, selenium and arsenic." In the waning days of the administration, the EPA "finalized a rule to limit what research it can use to craft public health protections, a move opponents argue is aimed at crippling the agency's ability to more aggressively regulate the nation's air and water." As president, Joe Biden undid these policy reversals, but in 2023, a U. S. Supreme Court ruling obligated his administration to weaken water protections.[37]

Water law matters. In 2014 in West Virginia, a tank farm owned by Freedom Industries fouled the perennial Elk River. After overwhelming the purification system of West Virginia American Water, 10,000 gallons of a chemical used in cleaning coal poisoned the drinking water of 300,000 people. Declaring a state of emergency, the governor told residents of a nine-county area, "Do not drink it. . . . Do not cook with it. Do not wash clothes in it. Do not take a bath in it." Forcing people to rely on water from elsewhere, the Freedom spill may have sickened 100,000, one-third of those affected, so it's no wonder that over two years later some still did not trust their tap water. Unfortunately, disasters like this one are neither isolated nor uncommon, even with laws in place to prevent them.[38]

Lackawanna River in Scranton, 1890.
(Courtesy of Lackawanna Historical Society)

Archbald pothole, 1938. *(Courtesy of Lackawanna Historical Society)*

Delaware and Hudson Coal Company gravity railroad and canal in Honesdale, 1898. *(Courtesy of Lackawanna Historical Society)*

Water pump underground, Olyphant colliery, 1921.
(Courtesy of Lackawanna Historical Society)

Scranton Lace Curtain Company, view from the Manville colliery breaker, with the West Ridge colliery breaker, left, and the Dickson colliery breaker, right, 1916. The Lackawanna River flows between the West Ridge colliery and Scranton Lace. *(Courtesy of Lackawanna Historical Society)*

Interior of wash house, showing locker room, Loree colliery, Larksville, 1919. *(Courtesy of Lackawanna Historical Society)*

Slush bank at "Y" at Jessup looking south, circa 1919. *(Used with permission of Pennsylvania Historical and Museum Commission, Anthracite Heritage Museum)*

Scranton Street, Scranton, August 1955. *(Courtesy of Lackawanna Historical Society)*

Dam #7, Scranton Gas and Water Company, 1911.
(Courtesy of Lackawanna Historical Society)

Marvine colliery washery and power plant, circa 1919.
(Courtesy of Lackawanna Historical Society)

Olyphant colliery with power plant, circa 1945.
(Courtesy of Lackawanna Historical Society)

Breaker boys, no date. *(Courtesy of Lackawanna Historical Society)*

Natural gas drilling site, northeastern Pennsylvania, circa 2010.
(Used with permission of Scranton Times-Tribune*)*

6

Rising Water

High water haunts humanity—so much so that flood myths worldwide make us all water-soaked survivors. People have told the Noah story for generations. Lamenting human wickedness, God decided to wipe the world empty but for Noah, his family, and pairs of animals, both clean and unclean. They all piled into an ark and rode waters covering the earth before landing in the mountains of Ararat. Pivoting on mass death, the tale points to a fresh start: God strikes a covenant with people that never again will a flood devastate the globe. He tells Noah and his descendants to "be fertile . . . and multiply; abound on earth and subdue it." We have heard, maybe too well, and the ark is now ablaze.[1]

High water frightens for good reason: it has a long history of killing people. The Yellow River has likely carried off more men, women, and children than any other place on the planet. In 1887, its flooding and the starvation that followed took as many as 2 million; in 1931, as many as 3.7 million. Not every rise of the river has been natural: to stop Japanese forces in 1938, a Chinese army blew up a levee, releasing a catastrophe of drowning, disease, and deprivation that killed 890,000. In the U.S. in 1927, storms along the Mississippi broke dikes, inundating millions of acres, displacing thousands of people, and killing maybe more than 1,000. To save New Orleans from high water, officials exploded a dike downstream, flooding places populated by the poor.[2]

Flooding at mines can touch off unforeseen effects, imperil miners, and increase company costs. In 2016 in Montana, migrating snow geese landed in an open pit copper mine filled with bad water. Thousands died. To keep from killing more birds and incurring more fines, the owners of the site have since frightened off flocks with "lasers, drones, fireworks and remote-controlled boats." The American Petroleum Institute argued against protections—and fines—by blaming the birds, which, it claimed, "are the actors, colliding or otherwise interacting with industrial structures." Apparently, geese should read the signs. To undo the forfeitures, in 2020, the Trump administration sought to end the criminal penalties attached to the 1918 Migratory Bird Treaty Act, despite industries annually killing 450 million to 1.1 billion birds, out of a total population of 7.2 billion in North America. In another reversal of Trump policies, in 2021, the Biden administration restored the penalties.[3]

High water has other spillover effects. A flood near Hazleton led to a coal strike that triggered a massacre. After a subsidence in 1877 stopped most mining in Harleigh No. 3, the adjoining Ebervale workings took on more water. When a fire broke out in Ebervale in 1885, workers plugged holes to keep Harleigh seepage from hampering efforts to kill the flames. Harleigh soon filled, but after the fire was out, Ebervale managers refused to accept any Harleigh water, despite pressure on the boundary pillar between the workings. With enough to pump, why should they take in Harleigh water? Just as stubborn and cost conscious, Harleigh managers refused to dewater their tunnels, despite knowing that a burst barrier might drown men in Ebervale.[4]

Worried about pillars, a mine inspector foresaw that a subsidence might suck a surface stream into Harleigh. Only after he forced a work stoppage did the owners of both operations agree to pump the minimum necessary and build a ditch to carry the brook beyond the mines. As work on the channel stalled, however, the rain-swollen creek poured into Harleigh. To stop the flow, workers threw into it what they could—hay, logs, stumps—but it all disappeared. Renewed rains defeated repeated efforts to staunch the flow, and both mines filled, making it too expensive to pump them dry.[5]

Millions of tons of anthracite were too tempting to leave unmined. In 1891, crews began carving Jeddo Tunnel, a seven-by-nine-foot opening cut from the floor of Butler Valley toward the summit of Broad Mountain. Completed as a drought dragged on in 1895, the tunnel drained not only Harleigh and Ebervale but also other mines north of Hazleton, including Lattimer. With the gravity system giving operators north of the city an economic edge, those south of town imposed miserly work requirements. Ordered now to spend two unpaid hours walking to and from a distant barn, mule tenders expressed their resentment at a mine foreman, a scuffle broke out, and soon a work stoppage spread, climaxing with 50 unarmed strikers dead or wounded in the Lattimer Massacre.[6]

Floods Underground

To prevent underground floods from spreading, in 1891, Pennsylvania required coal operators to maintain barrier pillars between workings. Although robbing other pillars was common knowledge and perfectly legal, the law intended boundaries to be untouchable. Despite this, companies often not only robbed them but also obtained permission to cut through them, especially when an operator owned adjacent mines. As the mine drainage problem deepened, engineers worried that water pressure in mine pools might break barriers, potentially causing a cascade of failing pillars.[7]

Underground floods could be deadly. With few owners shouldering responsibility for mines they abandoned, water sometimes surprised those who mistakenly tapped forgotten workings. In 1891, in the Grand Tunnel near Plymouth, miners dynamiting the face breached an unused mine. Water rushed down the gangway, men fleeing before it. Three who didn't find the surface may have escaped into other tunnels; if so, one paper speculated, "a siege of starvation and final death is ahead of them."[8]

On the same day, in Jeansville, near Hazleton, two workers unknowingly drilled into an old mine. As a score of men fled, water poured through tunnels, filling them to the surface in five minutes, a distance of 624 vertical feet. Observers initially assumed that 18 miners had drowned, but rescuers discovered 12 bodies showing no signs of drowning. Several of the dead had built a hut, lit a fire, and lived at least eight days before suffocating. A 13th man, who tried to swim to safety in the darkness, didn't make it. A full 20 days after the flood, rescuers happened upon 4 survivors who had lasted by eating blasting paper and drinking mine water and their own urine.[9]

As at Harleigh, subsidence sometimes flooded mines, putting operations on hold and adding to production costs. In 1899, two cave-ins exposed the Schooley mine in Exeter to the Susquehanna River. Torrents gushed inside in "alarming" volume, despite "strenuous measures" from crews manning pumps and a hoisting bucket. The company put masons to work building underground dams, but 10 days later, seven pumps and two hoists could not stay ahead of the flood. After another several days, a mine cave suspended pumping; engineers feared an explosion if the squeeze had damaged the boiler house. A month after clearing began, the effort ended in defeat. From then on, officials could only watch the water rise.[10]

In 1924, a mine cave in Scranton swallowed the Lackawanna River. Rushing below at 600,000 gallons per minute, the water killed two workers, knocked out pumps, and inundated five mines, putting 7,000 out of work. Desperate to stop the flow, men tossed in mine props, mine cars, and tons of rock. All disappeared. They then dropped into the hole two 40-foot steel beams, tossed on top a steel lattice, and dumped onto the web a train of loaded boxcars. Hours later, after the river slipped back into its former bed, the crews turned to pumping 17 million gallons daily from the tunnels, a task projected to take two months.[11]

Mine owners flooded mines on purpose, often to snuff a fire but sometimes to hide shady operations. In 1891, a blast at the face of the Mills seam in Nanticoke ignited gases that set the tunnel ablaze; after conventional efforts failed to kill the fire, the order came down to flood the mine with 15 million gallons of water. Ten months later, the tunnel still lay idle. In 1902, Scranton Coal Company siphoned water from the Lackawanna River to its Richmond No. 3 colliery to extinguish a fire; filling the workings took several weeks, followed by an expensive, two-year pumping operation. In 1920, Peoples Coal Company shut down pumps protecting its mines in West Scranton to hide from city officials all evidence that its mining practices caused costly cave-ins.[12]

Knox Mine Disaster

Flooding imperiled the industry itself. Engineers had long warned that the buried valley of the Susquehanna posed a threat, but companies paid scant attention. In 1950, a federal report about the unseen danger again sounded the alarm, noting that "if, because of subsidence, a cave should occur and water from the Susquehanna River flow suddenly into the mine workings, a major catastrophe could result." Nine years later, it did. In January 1959, men in the River Slope tunnels of the Knox mine near Pittston chipped too close to the old riverbed. An icy Susquehanna burst in, flooding gangways and killing 12.[13]

The subsidence spawned a whirlpool. To close the hole, workers diverted nearby train tracks and rolled 50 coal hoppers and over 400 mine cars into the maelstrom. When a construction crane dropped in a 30′ × 30′ steel mesh like that used in 1924, the vortex undid it. Meanwhile, laborers also dumped in loads of culm, boulders, and wood shavings, all to no effect. On the third day, finally, the swirl slowed, and soon the gap sealed. Over 10 billion gallons had poured into the opening; the underground mine pool extended three and a half miles by one-half mile.[14]

The dewatering job was of "a magnitude unseen in anthracite mining history." Before pumps could be installed, however, men wrestled from the mine "hundreds of tons of ice chunks." By February 3, 2 pumps pulled only 3,000 to 6,000 gallons per minute from the River Slope and Schooley shafts, but by March 17, 40 pumps sucked up 142,000 gallons per minute, eventually drawing out 11 billion gallons, a billion more than the Susquehanna had poured in. After the mines cleared, crews constructed a cofferdam, drilled over 100 boreholes, and sealed the breach with 1,800 cubic yards of concrete and 26,000 cubic yards of sand.[15]

Although observers at first attributed the catastrophe to "freakish weather" and a rising river, formal investigation confirmed that company owners knowingly kept workers mining too close to the buried valley. Typically, companies left 30–80 feet of rock between underground workings and the bottom of the old river, but when Knox operators extended tunnels under the Susquehanna, they redrew maps and narrowed this support in places to as little as 19 inches. At the point of collapse, the distance may have been only 6–8 feet. Adding to the scandal, no one conducted borehole tests for the location of the buried valley, and the District 1 United Mine Workers president secretly served as a co-owner of the business, offering evidence that across the region, the underworld of organized crime and the upper world of "normative order" often worked in "close cooperation." Courts convicted a dozen people and three companies of bribery, income tax evasion, and labor-law violations.[16]

The flood all but ended mining in the Wyoming Valley. Dewatering not only emptied Knox mines but also revealed the costs of mining, which large companies now refused to shoulder. Not long after the disaster, Pennsylvania Coal Company and Lehigh Valley Coal Company left the hard coal business. With local unemployment already over 11 percent and the median income only 80 percent of the state average, the regional economy shuddered at the loss of another 7,500 jobs and $32 million in payroll. Compounding the pain, towns feared lost tax revenues, and a sharp uptick in mine subsidence placed yet another burden on communities across the valley.[17]

In the Knox disaster, the feedback loop at the center of the water-energy nexus snapped. Happening at a pivotal point—just as the costs of mining water outstripped the profits from mining coal—the flood placed a period at the end of deep mining in the Wyoming Valley. With anthracite sales unable to pay the price for pumping and with so many tunnels suddenly waterlogged, the industry was all but dead.

Similar floods have happened across the world. In 1915, without warning, ground gave way under the Sea of Japan and "water rushed into the gap, churning angrily," drowning 237 miners at the Higashimisome colliery. After the disaster, pumping crews noticed that the level inside, 147 feet beneath the sea, moved in sync with ocean tides. In 1995 in India, mine fires plaguing the Gaslitand mine may have weakened coal pillars enough that monsoon rains triggered a cave-in that diverted a waterway into the tunnels, drowning 74 workers. Eroding the financial health of the company, the damage killed any chance that it could earn enough money to fight the fires.[18]

In the twilight of hard coal, the Conowingo tunnel project and the Knox pumping operation highlight the efforts that the state expended to prop up the anthracite industry, which floated what remained of the local economy. As flooding increases today due to climate change, federal and state governments similarly invest time and money to protect the infrastructure of the fossil fuel industry. Petroleum companies want the federal government—taxpayers—to build a "nearly 60-mile 'spine' of concrete seawalls, earthen barriers, floating gates and steel levees on the Texas Gulf Coast." The $12 billion project would shield "one of the world's largest concentrations of petro-chemical facilities, including most of Texas' 30 refineries, which represent 30 percent of the nation's refining capacity." After Hurricane Harvey devastated Houston in 2017, Washington put up nearly $4 billion to get the work started. A strong enough hurricane, though, could still flood the Houston ship channel, despite piles of earth, concrete, and steel. Texas politicians, some of whom deny climate change, help funnel billions of taxpayer dollars into the pockets of the fossil fuel industry: "between 1950 and 2010 the US government had given $369 billion to oil companies, $121 billion to natural gas companies, and $104 billion to coal companies."[19]

Lackawanna River Floods

Anthracite and water have a long history here of not mixing well. From the earliest days of mining in the Lackawanna Valley, high water disrupted coal operations. Even before a Wurts crew sank the first shaft in the valley, the Lackawanna River spilled into the brothers' open pit diggings, swelling high enough in 1830 to fill quarries on both banks. Rainfall later burst the dam at a D & H reservoir above Carbondale, sending a torrent through town, filling mines, and drowning two underground workers. Sometime later, another badly built dam unleashed a wave that destroyed more miners' homes and killed as many as eight.[20]

The "Lack-of-water" usually flowed with little force, but three major floods, one after another, struck the valley just before and after the Anthracite Coal Strike of 1902. The first hit in December 1901, a second in February 1902, the third in October 1903. The high water called renewed attention to the river's narrowing channel and its rising bed, due mainly to the spread of washeries, which had recently become important to company bottom lines.[21]

Each a record breaker, the first two floods forced people from their homes, inundated mines, and washed out bridges and railroad tracks. In the 1902 rise, the Briggs washery tipped over, men dynamited ice that threatened reservoir dams, and water overwhelmed homes across the city, signaling that new storm water systems could not cope with runoff. Observers thought that the 1902 high-water mark would "not likely to be reached again, perhaps for generations," but the next year, the river shattered its own record.[22]

The 1903 flood rose from over five inches of rain falling in 48 hours. Reaching 8 feet above its 1902 crest, the Lackawanna flooded nine collieries, dismantled dozens of bridges, and made 250 families homeless. As men and mules struggled to escape from mines, water poured down shafts, flooding engine rooms and overwhelming pumps. Touching every colliery near the river, the high water trapped three men in a mine in Mayfield; in Olyphant, crews blew up riverbanks to open a channel; and in Carbondale, where water rose 10 feet beyond previous records, the Lackawanna ripped out D & H tracks, sweeping away 30 loaded coal cars. After the flood had washed hundreds of tons of coal into the stream, people all along the river combed its bed for burnable anthracite.[23]

Many blamed all three disasters on companies that "had filled the river bed so much that in places it was higher than the banks." After the 1902 flood scoured sites where they had dumped waste, the same outfits simply continued the practice, among them Scranton Gas and Water Company. By 1903, a culm pile of the Scranton Coal Company had reduced the stream from 150 to 50 feet wide. Although a narrowed channel and increased flooding made for "not a place in the city where it [was] safe to build near the river," people still crowded homes and businesses on its banks.[24]

Scranton officials knew all about the changes, but they could do little. In 1892, they sued a steel company for filling the channel, only to fail on appeal to the state supreme court. In 1899, a city engineer reported that culm and company waste had altered the river in places as much as 100 feet from its course in 1857. After the 1903 flood, the city decided to ask companies to stop dumping and to remove ashes and culm from the riverbed. If officials made the request, it went unanswered, perhaps because ordinances against dumping had long before become a "standing joke" among offenders.[25]

One could see that ashes tossed from riverbanks narrowed the channel; more difficult was noticing that culm washed from waste banks and washeries raised the bed. The problem made flooding within the city worse, but the solution lay outside of town. All along the Lackawanna, from Forest City to Pittston, "companies [were] daily running off tons of fine culm into the river." Too shallow and too slow, the stream could not flush enough away fast enough, leaving waste to fill former swimming and diving holes. Choking the channel, culm deposits slackened its flow, making it even harder for water to carry off the coal, which led to ever deeper deposits. This feedback loop sharpened the "unhealthy odor" of the river and created its "appearance of filthy, stagnant water."[26]

A lot of coal washed into the Lackawanna and Susquehanna Rivers, raising their beds, multiplying flood risks. In 1902, water pumped from mines and drained from washeries increased the flow of the Susquehanna by about 20 percent. In 1903, at the confluence of the rivers, culm extended into the Susquehanna "several hundred feet in length and nearly as broad, and . . . fifteen or twenty feet [deep] . . . under the water." Although culm piles were "the worst offenders," washeries added waste to the Lackawanna to the point where "it takes but little more than an ordinary rain to cause the river to overflow." In 1944, engineers estimated that in dry times, mine water made up 50 percent of the flow in the Lackawanna. Collieries in both basins poured into the Susquehanna a combined "average of 491.9 cubic feet of waste per second," altering the river's course, forming islands, and adding ground to others.[27]

To bend the Lackawanna to human will, people clamored for flood control. Scranton's future prosperity, a neighborhood board of trade argued, depended on the city building walls "to make the river what it should be, a stream that will always have certain bounds and [be] fully able to meet the cleansing demands made upon it." To protect its railroad, the D & H dug through 13 feet of culm to build a barrier against flooding, a dike that simply passed high water downstream. Still the river pushed back.[28]

Enough rain can overwhelm. In mid-March 1936, the Lackawanna inundated mines in Duryea and Pittston, throwing as many as 7,000 men out of work. Finding a mine cave on its banks, the river widened the gash to 300 feet and cut 39 feet deep to reach the upper veins; floodwaters poured below for 72 hours before crews could divert the current. The high water revealed that in planning flood control, the federal government had never considered how to prevent the river from inundating mines. Immediately after the disaster, U.S. Army engineers estimated that it would take up to a year to pump the tunnels dry. The flood was costly: in addition to lost production time, mining companies paid workers nearly $100,000 for missing tools. In places in the Wyoming Valley, the Susquehanna rose a foot higher than its 1902 crest. Contributing to passage of the national Flood Control Act, the damage in the Wyoming Valley led to the construction of levees along the river in the 1940s.[29]

Other floods followed the 1936 peak of 33 feet, which topped the 1902 rise at Wilkes-Barre of 31: 1972, 41 feet; 1996, 34 feet; 2004, 35 feet; 2011, 42 feet. More encroachment, higher population, expanded development, further channelization, and increased rainfall combined to keep floods coming. The lesson for a warming world: high water prompts control, which leads to higher water, which causes more destruction, which prompts more control, which triggers more floods.[30]

Diane and Agnes Floods

The average annual rainfall in northeastern Pennsylvania is 38 inches. In 2018, record rain fell: 61 inches. The next year proved not so wet, but rains still totaled 49 inches, 11 above average. If a lot of rain spreads across 12 months, the land and people usually can handle it, but when it comes at once, it often does serious damage, a problem that climate change will worsen.[31]

For every degree Celsius rise in atmospheric temperature, the amount of moisture in the heavens expands by 7 percent. Said another way, for every degree Fahrenheit rise, the atmosphere absorbs and dumps 4 percent more water. Downpours already happen more often: in the Northeast, the heaviest rain events from 1958 to 2012 increased in intensity by 71 percent. With more water in the air, snowstorms will also intensify, swelling spring melts. Amplifying water worries, climate change has altered the jet stream, slowing storms and increasing precipitation.[32]

Disturbing few locals, rain from Hurricane Diane fell off and on during August 18, 1955. Then the skies unloaded, in under an hour raising streams 30 feet. Near Stroudsburg, water carried away more than 36 campers; officials feared the death toll would climb to near 80. At 88 places, torrents tore out track on the Lackawanna Railroad, requiring "a virtual rebuilding . . . across the Pocono mountains." In Scranton, rain goaded Roaring Brook to rip a hole 1,000 feet long and 60 feet deep, destroying rails at the Nay Aug tunnels. Not far on, the stream erased the Little England and Flats neighborhoods; several square miles of these communities were never rebuilt. Diane inflicted damage that neither the Lackawanna nor Erie Railroads could afford to repair, forcing the two to merge. Lessons from the storm went unlearned, however. After high water in 1972, the Erie-Lackawanna Railroad declared bankruptcy, practically ending railroading in the region.[33]

Hurricane Diane also pummeled Hazleton. When workers expanded the Jeddo Tunnel system in the 1930s, they carved out a pump room 860 feet beneath the surface to suck water from deeper regions. Catching crews off guard, Diane sent men scrambling to close the doors to the pumps, but water stopped them and the mines filled. The result: an accelerated decline of local coal mining.[34]

In the late 1930s, people dismissed the idea that the 1936 flood could be repeated, an attitude similarly expressed after the 1902 rise. Too many placed too much faith in flood control systems to keep at bay the next crest of the Susquehanna. For several days in June 1972, Tropical Storm Agnes swirled over the region, dropping about 14 inches of rain in one 24-hour stretch, exploding the Lackawanna and Susquehanna Rivers, topping levees, and devastating Wilkes-Barre. Covering 60 percent of the city, floodwaters not only coursed through mine tunnels but also unburied 2,500 dead in the Forty Fort Cemetery, scattering body parts to turn up on "back porches, in basements, on roofs." At the time, Agnes registered as the costliest natural disaster in U.S. history, with over $3 billion in damages; destruction in Pennsylvania accounted for over $2 billion, making it the worst in state history. The death toll: 117.[35]

The damage that Agnes inflicted reveals the trouble with levees: they try to control what can't be tamed. In 1973, one observer calculated that preventing a similar catastrophe would require 522 new dams in the Susquehanna River basin; building them, though, would be "at best impractical and at their worst potentially more costly than floods themselves." On the other hand, floodplains have long mitigated damage on major rivers by letting high water spread out, rather than roar downstream to overwhelm unguarded places. Remembering the trauma of Agnes should have led to fewer artificial controls, more attention to floodplains, and the Wyoming Valley getting right with water. Instead, levees went higher.[36]

"Stay Home"

Getting right with water means finding a middle ground between interference and inaction, excess and lack. Poets can help because they question everyday definitions, interrupt our usual thinking in surprising ways, and remind us of possibilities that we had overlooked. In a world confronting climate change, we need poets to define our work from fresh angles to know better how water hurts and enhances home.

Reading is loopy, a fact that Robert Frost expresses when he says that to know a poem is to know other poems. Reading, he claims, is about circulation, not progression: one must "get among the poems where they hold each other apart in their places as the stars do." About observation and reflection, a good poem refreshes how I see, clarifying my vision, which reading a second poem often revises, sharpening my sense of the first poem—and the world. I experience this loop when I return to "Stay Home," Wendell Berry's response to Frost's "The Pasture." Berry's poem leads me to reinterpret Frost's, which gets me to rethink Berry's. A conversation, the poems express different visions of how to work with water.[37]

"Stay Home" does not mean staying put. Answering the invitation in "The Pasture" to help with spring farm chores, the speaker in "Stay Home" does not so much refuse the requests as call attention to nature's taking care. The poem defines "labor of the fields" (4) as not only human work but also as the work of the field itself, which grows grass, food for the cow in "The Pasture." About protection, the word "home" in "I am at home" (6, 13) links land and being: "I" exist in a place that I care for and that cares for me. In asserting "Don't come with me" (6, 13), the speaker encourages the reader not to mimic him, but instead to "stay home" (7, 14), to be somewhere, in the sense not only of taking up residence but also of providing support for, a deeper commitment to place than an address or a zip code.[38]

In nodding to different parts of the water cycle, "The Pasture" and "Stay Home" debate the nature of work. Water flows in the first stanza of each; the second stanza describes what water gives rise to. In "The Pasture," humans corral water welling from the earth; they intervene. In "Stay Home," water falls freely from above; the speaker waits, letting rain act on its own. "The Pasture" represents how the flow of a spring—inspiration—each year brings forth new life, whether calves or poems. Pointing to perennials, grass and trees, "Stay Home" remembers the long view, cautioning the reader to watch and wait before acting or writing. To do responsible work, both poems agree, a farmer must study the land, know its waters, and then labor, which is to act in accord with the place. Only then does annually cleaning a spring make sense; only then does farming—sustaining life for the long haul—make sense. Feedback loops, these acts nourish a human presence in the world without destroying the world. Mining does the opposite.

Politicians also play with language, too often to disastrous effect. In 2002, the Bush administration rewrote the Clean Water Act when it redefined the term "fill material" to include toxic mine waste, greenlighting coal companies to push remnants of mountains into surrounding valleys. Mine owners in the southern Appalachians let the fill grow into earthen dams that held back millions of gallons of wastewater. Undermining clean-coal narratives, these slurry ponds have sometimes failed spectacularly.[39]

Prior to the 2010 BP oil disaster in the Gulf of Mexico, the largest toxic spill in U.S. history was the 2000 collapse of a coal waste pond in Martin County, Kentucky. Slurry flooded abandoned mines, pushed through tunnels, and surfaced to choke 100 miles of streams. In Tennessee in 2008, a coal ash pond at Kingston Fossil Plant broke, burying 15 homes and 400 acres in about six feet of sludge. Similar episodes have happened elsewhere. In 2019, a dam in Brazil failed, freeing mine mud that overwhelmed 166 and left 147 missing. The catastrophe echoed a similar dam collapse in 2015 at the same company's Mariana mine, which killed 19, left 250,000 without drinking water, and flushed 60 million cubic feet of waste into rivers and the Atlantic Ocean.[40]

A flood alters space and time. Overwhelming boundaries and refusing limits, it briefly remakes a place into a moving surface singleness, beneath which stretches a former world, now new, ready for revelation when the water recedes. As it does, deposits of mud and debris, torn from other places, other lives, crowd familiar sites. The rewriting not only triggers a forgetting but also invites a remembering. Regaining wholeness, however, requires extra care when, as here, flooding resurrects a toxic past.

7

Damming Water

Dams have transformed lives and lands across the globe. After World War II, construction escalated, uprooting "at least 80 million rural people," with the Three Gorges Dam alone unsettling 1.3 million in China. Behind this boom stood World Bank loans, which backed building thousands of high dams—50 feet and more. Although big dams look like easy answers to water and energy demands, they displace too many, cost too much, and too often damage ecosystems. Recent widespread opposition has pressed governments and banks to shy away from them, despite their hydropower promise.[1]

Pushback extends to their danger: dams can collapse. The world's worst recorded dam disaster happened in 1975 when a typhoon ripped out two in central China, releasing a wave 20 feet high and seven miles wide, killing between 80,000 and 200,000. In 2020 in Michigan, one on the Tittabawassee River failed, overwhelming a second, forcing 10,000 to evacuate. Two dams came apart in 2023 in Libya, sweeping away thousands. Heavy rains due to climate change may imperil more; storms cracked big dams in the U.S. in 2015, 2016, and 2017.[2]

In hard coal country, drought raised dams. In 1895, a rainless spring-to-fall period fired competition for water among local cities, collieries, and water companies. The dry spell spurred dam building, triggered legal battles between coal and water enterprises, and forced water businesses to merge. Revealing vulnerabilities in the water-energy nexus, the drought further empowered monopolies of coal and water, reshaping water cycles across the Anthracite Region.

Private versus Public Water

Water connects all in its circulations, making it inviolable for many. But industrialization separates land and water to treat each as "an abstract commodity," sullying both, transforming water from sacred to exploitable. Seeing water only as a resource robs it of life; no longer a gift, it becomes just a thing, often a tool. Reducing it to input, instrument, or product makes it available only to those who can pay. The poor may go thirsty. As technologies advance, climate change accelerates, and droughts recur more often, worries rise over how to know water.[3]

Tensions turn first on whether water is a private or public good. Scarcity puts a price tag on most things, so debates about privatizing it take on urgency in places threatened with longer and more frequent droughts. Economists claim that pricing saves water, just as it does oil, because the more that people pay for something, the more likely they will use it wisely. But if a Texan runs out of oil, his truck will die; if he runs out of water, he will die. As with oil, selling water means withholding it for payment, but with good water necessary for life, not just the economy, it should be available to all. Ensuring access, though, costs money for pumps and pipes.[4]

To pay for them, many powerful people would prefer to treat water like oil. In 1992, the International Conference on Water and the Environment asserted as a guiding principle that "water has an economic value in all its competing uses and should be recognized as an economic good." Viewed this way, it requires institutional management that relies on "incentives, prices and markets and less reliance on traditional command and control approaches." In 2008, oil tycoon T. Boone Pickens bet on water as the new oil when he described plans to supply Dallas from the depleting Ogallala aquifer. In 2012 in the desert city of Dubai, the Global Water: Oil and Gas Summit brought together "representatives from Shell, Saudi Aramco and Conoco-Phillips" to figure out how to sell wastewater to the public; attendees decided that putting a price on fresh water would make used water more palatable, bestowing on it "higher worth in an era of water scarcity."[5]

Others argue that water is no commodity. In 2000, shortly after Cochabamba, Bolivia, privatized its water system, people protested the change, declaring that "water is a fundamental human right and a public trust to be guarded by all levels of government, therefore, it should not be commodified, privatized or traded for commercial purposes." In Michigan in 2003, the Earth Liberation Front planted four bombs at a Nestlé pumping station; in taking responsibility for the explosives, which did not detonate, the group proclaimed that "clean water is one of the most fundamental necessities and no one can be allowed to privatize it, commodify it, and try and sell it back to us." In 2002, the United Nations Committee on Economic, Social, and Cultural Rights asserted that the right to water is a human right. It should go without saying that people have a just claim to water, but in 2010, the UN General Assembly passed a resolution stating that humans do, in fact, have this right. In a sign that the debate continues, in his 2015 encyclical *Laudato Si'*, Pope Francis pointed to arguments to privatize water, and in 2017, he signed the Rome Declaration of the Human Right to Water and Sanitation.[6]

Especially in a climate change context, water raises questions about interdependence, justice, and responsibility. A leveler of classes—everyone needs it—water exposes economic and social anxieties over its availability, purity, and use. With the earth growing hotter, more crowded, and more urban, no one can simply assume access to clean drinking water. In a few years, almost two billion people will suffer "acute water scarcity and two-thirds of the world's population will live under water stress." Add to this the specter of drought and tensions multiply. As the Anthracite Region discovered in 1895, droughts force choices about who has rights to what water.[7]

Drought and Mining Coal

Floods frighten people, but droughts scare them more. Although classical civilizations probably thought that they had a right relationship with water, dry times ended several, in Egypt, Mesopotamia, Greece, Israel, India, Afghanistan, China, and Central America. In 2006–2010, record-breaking drought in Syria displaced farmers, inciting a civil war that killed thousands. In 2018, Cape Town barely escaped Day Zero, the moment when the city would go dry. In the Czech Republic in 2018, the Elbe River dropped low enough to reveal "hunger stones," rocks that warn people of tough times ahead. One stone, "considered the oldest hydrological landmark in Central Europe, bears a chiseled inscription in German that says: 'When you see me, cry.'"[8]

With history telling me that water sources can dry up, I pay attention when I hear that global aquifers are now emptying faster than first assumed. As California endured drought in 2011–2019, irrigated land in the state sank due to water withdrawals. In 2023, Arizona halted home building near Phoenix in areas that rely on groundwater "because there wasn't enough water for the homes that had already been approved." Beneath the Great Plains, the Ogallala aquifer has been drained to the point that pumping costs are "driving much irrigated agriculture in the region out of production." Much like coal mining, water mining is a "one-time extraction . . . from a depletable reserve." Unlike coal, however, fossil water has no substitute.[9]

A feedback loop: places forecasted to receive more rain due to climate change will also experience more droughts because warmer temperatures increase evaporation. Pennsylvania may feel the heat. By 2035, the temperature in the Northeast will have risen more than 3.6 degrees Fahrenheit beyond preindustrial averages, the highest hike in the lower 48 and about 20 years before global averages hit the same mark. Accelerating cycles of drought and flood will hamper natural gas mining in the state, currently the nation's second-largest producer, because drilling demands a steady stream of water.[10]

Pennsylvanians tend not to remember that dry times disrupted coal mining. Defined then as the "greatest drought this country has experienced in the last 100 years," a dry spell in 1895 stretched across the commonwealth from April to November, devastating the middle and southern anthracite fields. In all four coal regions, the dryness imperiled mining, upset markets, and exposed stresses between coal and water businesses. For some anthracite operators, water was the new coal.[11]

Lack of rain affected mines across the East and Midwest. In Indiana, Ohio, West Virginia, and western Pennsylvania, the drought dropped river levels too low to carry coal, idled 8,000 miners near Pittsburgh, and stranded barges on the Monongahela River with 20 million bushels aboard. By end of October, Pennsylvania Railroad ran dozens of water trains along its route between Philadelphia and Pittsburgh because reservoirs along the line had gone dry. The railroad also transported trainloads of water to its locomotive works in Altoona, which otherwise faced closure and the layoff of 5,000. As the Silver Creek and Beechwood mines near Pottsville shut down, collieries close by limped along, dependent on water trains.[12]

Touching the entire Anthracite Region, the drought drew awareness to how collieries interrupted natural water flows. Lack of rain aggravated the situation on Wolf Creek, a major feeder of the Pottsville reservoir. Six years prior, the stream measured 12 feet across and 18 inches deep; by 1895, a 12-inch pipe accommodated its inch-and-a-half flow. Blaming this contraction on extensive timber cutting and the spread of mining, the *Pottsville Miners Journal* called attention to the "vast quantity of water" that collieries consumed, which did "more to destroy the streams than any other cause"; a water company official explained that the "new process of washing coal makes it necessary to use rivers of water, while the extension of underground operations absorbs many of the streams."[13]

As the drought intensified, its effects worsened. Mine operators who found it "exceedingly difficult to secure water for steam-making purposes" turned to using "sulphur water." Acidic mine water required adding lime before feeding it to boilers, but the mixture still caused machinery to wear out. As even this water dried up, more collieries closed, although some in the middle fields turned to local water companies. In Carbondale, D & H workers retrofitted "old gravity boilers" to carry water to steam engines of the gravity railroad. Low water in early September halted traffic on the D & H canal, despite the ditch's extensive reservoir system. By mid-September, several towns in the Lackawanna Valley faced shortages that required hauling water by wagon from elsewhere. In Wilkes-Barre, a water company rescued another that could not supply customers. By late September, trolley cars stopped in Hazleton, industries and collieries closed, and 10,000 men lost work. In early October, communities near Shamokin and Mount Carmel announced that they had water for only a few days more.[14]

Lack of water upset coal markets. D & H production dropped by a third. With less anthracite mined, prices shot up in September, the wholesale cost jumping in a few weeks by 75 cents a ton. With coal supplies tightening just ahead of winter, prices spiked again in October. Broken and nut sizes were "not to be had—not a pound. The drought is at the bottom of the business. The mines can't be operated for the want of water."[15]

Not all coal companies suffered equally. Although many mines in the southern and middle fields suspended work, collieries of the Delaware, Lackawanna, and Western Railroad broke production records. In Scranton, operators had access to "abundant water" from Scranton Gas and Water, which called for no restrictions on use due to its large-capacity reservoirs. A flood dampened spirits at the D. L. W. repair yard, however, when the dam at the ironworks burst, sending a wall of water into the neighborhood of South Side. November rains ended the drought; closed collieries in the Lehigh region resumed mining on November 4.[16]

The drought prompted dam building, a technological fix. To cope with the lack of rain, Pottsville and Shamokin water companies planned new reservoirs. Backers of a proposed water company sought capital to raise dams to supply mines and railroads in the middle coalfields, but when investors contemplated buying rights to the headwaters of the Lehigh River, Lehigh Coal and Navigation (LCN) rejected the idea. Further irritating LCN, Wilkes-Barre water companies also eyed the upper Lehigh. Meanwhile, at the headwaters of the Lackawanna, plans for a dam raised a ruckus when Valley Water Supply Company moved to seize five acres of D & H land.[17]

The Water Knife and Faith in Technology

Some who wonder what climate change futures might look like turn to science fiction. Set in the near future, Paolo Bacigalupi's novel *The Water Knife*

(2015) dramatizes what could happen as the world careens toward climate collapse. Ending an everyday life that many take for granted—clean water, clean clothes, and a clean body (176–177)—a long drought grips the American Southwest. Without water, Phoenix shrivels, Chinese companies raise arcologies in the ruins, and an epic humanitarian crisis grows, sparking anarchy, internal migration, and class warfare. As the violence deepens, "everything about it was a feedback loop, building itself into something bigger and more horrifying" (35). Watching the world undone, I realize that to know climate change, I need to see with new eyes.[18]

Blind faith in technological fixes can obscure an existential threat. To escape mega-drought, the wealthy in *The Water Knife* retreat to arcologies, self-sufficient, low-impact, high-population buildings. Each a "living machine" (91), they may combine architecture and ecology to escape drought, but they are only "bigger and more ambitious than all of Sin City's previous forays into the fantastical combined" (10). Symbolizing their riskiness and priorities, the arcologies operate ground-floor casinos (4).

Hard and soft infrastructure, arcologies manufacture a wet world. Temperate zones, each concrete-and-steel tower circulates and recycles water through plants and fish that have been "carefully engineered and selected to accomplish specific cleaning tasks" (185). Advertising "clean air" (89) and "carefully constructed serenity" (173), a Phoenix arcology offers residents "a space that felt as if it were entirely removed from the dust and smoke and collapse of the greater city beyond" (172).

Arcology self-sufficiency and safety are illusions. To feed its greenhouses, the Taiyang Arcology in Phoenix must collect human waste from the city (94); the Cypress in Las Vegas can hold out for only three months before recharging its water supply (52). Although security systems at the Taiyang "controlled its borders as rigorously as Nevada or California" controlled theirs (172), a hired gun finds his way into its residential area, discovers two people murdered, and escapes to the street through the water treatment system. He knows the violence behind every arcology because he did the dirty work that built them in Las Vegas (8): forcing farmers to sign over water rights, crippling the Central Arizona Project, and organizing militias to stop climate refugees at Nevada's borders.

Technology not only lulls people into a false sense of security but also shatters it. Taiyang residents see their vulnerability reflected in the TV spectacle of a dam bursting. At least one watcher knows it was no accident: "The girl was staring up at the broken dam. 'Who did that?'" (206). The looping news clip stuns viewers not only because it depicts the undoing of social taboos against wasting water during a drought but more because they experience a superabundance of water out of human control. Overwhelming all it touches, the flood reveals the fragility of dams, despite their concrete and massive size.

D & H Drought Fears

Water always worried the D & H. After crews finished constructing the canal in 1826—"during a season of unusual drought"—workers discovered that it leaked; to become impermeable, its sides needed saturation and sufficient time to settle. From its start, the company feared that too much water might flow to other businesses; in 1827, it suspected that the enterprising Jason Torrey would divert water to his sawmill. When Torrey petitioned the company in 1840 to siphon water from a D & H dam, president John Wurts wrote back to say that "no saving of money and no feeling of friendship or kindness would induce us to run any risk on the score of water. . . . We must act not for a day or a year, but for all future time." Guarding its right to water, the D & H had similar brushes with others. In 1838, its board "warned mill owners at Eddyville [New York] that it would take whatever measures were necessary to guarantee a water supply if they did not repair their dam." Adding to its anxiety, the D & H warred against what it could not sue: eels, catfish, and moles, whose diggings contributed to keeping the company in the red until 1840. Water also demanded routine attention: leaks had to be repaired, water levels had to be monitored, and each year the canal had to be filled in early spring and drained in late fall.[19]

Enlargement of the canal in the 1840s and drought in the 1850s only increased water worries. The combination forced the D & H to invest in buying ponds and building reservoirs. With every opening of a lock—110 total—releasing downstream about 100,000 gallons, expansion meant more water. In 1846, the company purchased three ponds, two near Honesdale and one in New York. In 1852, drought forced the canal to rely on stored water from mid-June to early November. The next year, the D & H freely used its reservoirs, confident that it had an "abundance of water under control." Abundance, maybe; control, no. In 1854, lack of water endangered operations enough that the company had to lighten boats.[20]

Drought led the D & H to replumb watersheds, and by 1880, its workmen had constructed 22 reservoirs. In Wayne County, they created White Oak, Long Pond, Elk Lake, Mud Pond, and Swamp. Belmont Lake, the origin of the west branch of the Lackawaxen River, also fed water to the canal, along with reservoirs at Hankins and Miller Ponds, whose bluestone dams had flashboards that workers used to control releases. By damming wetlands, the company not only determined the flow of water to the Lackawaxen but also altered local ecosystems. These human-made, nature-culture spaces soon became "natural" sites for fish, plants, migratory birds, ice harvesters, anglers, and boaters. Its long fear of losing water provoked the D & H to push back hard against Valley Water Supply Company during the 1895 drought.[21]

The Water Knife and Required Reading

The Water Knife nudges me to watch my words because they define what I see: "Weather anchors used the word *drought*, but *drought* implied that *drought* could end; it was a passing event, not the status quo" (27). A water official points out that "our own words make us blind"; without the "right words in our vocabularies," she says, "we can't even see the things that are right in front of our faces" (59). After listing events that she didn't see coming—no snowpack in the Rockies, dust storms covering solar panels, and "California putting all these calls on the river" (59–60)—she laments that it's as if "someone is always ahead of us. Someone who sees more clearly than we do. Someone who has better words to describe where we're headed" (60). The novel ends with her poised to use legal language to cut off all water to Phoenix, creating millions more migrants as the mega-drought drags on.[22]

Replumbing places—installing larger pumps, higher levees, and taller dams—signifies human attempts to control nature. Recounting real-world efforts to master water in the American West, *Cadillac Desert*, "the bible when it comes to water" (181), explains that cities and farms arose in irrigated deserts based on the fiction of limitless water. The novel's Southern Nevada Water Authority requires employees to read this history so that they "see this mess isn't an accident" (160; see 164–165). Despite long knowing that drought could destroy the human presence in the Southwest, Americans not only "didn't do anything about it" (160) but also flocked to cities like Phoenix and Las Vegas, making them ever more vulnerable. A hydrologist who hides the most senior water rights inside *Cadillac Desert* tells a climate refugee that the book describes the "beginning of everything. When we thought we could make deserts bloom, and the water would always be there for us. When we thought we could move rivers and control water instead of it controlling us" (181).

In seeking to master nature, rather than to live within its rhythms, too many accept the fiction of limitless water, which another fiction underwrites: that one can claim alienable rights to water. Hiding water rights inside *Cadillac Desert*—hiding a legal fiction inside a history of irrigation in a novel about climate change—warns us to be wary of words, which can become untethered from what they define. "Desert" does not mean "garden." Heeding *Cadillac Desert* at its publication may have avoided the disaster that unfolds in the novel, just as heeding climate science now might mitigate the disaster that is unfolding today in the real world.

Lackawanna Water Grab

Conflicts over water increased in step with population and industrial growth in the Lackawanna Valley. As corporations competed for access to the same sources, dams rose as assertions of control, not only of streams but also of people and places. Sharpening the contest, the 1895 drought unveiled the uneasy interdependence of coal and water companies, a fact that sometimes manifested in overlapping ownership but most in the fraught lives of miners and their families. Whoever controlled Stillwater Gap determined the flow of the river, which meant influencing floods, pollution, and distributions of water for drinking, mining, and manufacturing.

At the northern tip of the Anthracite Region, Stillwater Gap joined the Lackawanna Valley to the Alleghany Plateau. In the notch, the Jefferson branch of the Erie Railroad, opened in 1870, ran tight against the Lackawanna and had to cross the stream twice as it passed through the ravine. In 1890, lack of room in the cut forced the Ontario and Western Railroad (O & W) to lay its line right alongside its rival's as they left the gap. The Erie then followed the west stem of the river; the O & W paralleled the east branch.[23]

Tensions over Stillwater stretched back at least to the late 1880s. Chartered in 1887, Valley Water Supply Company looked to erect a dam at Stillwater to compete with Scranton Gas and Water Company, the city's main supplier. Not far upstream from Forest City, the dam would be located just beyond the coal measures, ensuring that acid and culm could not contaminate the reservoir. The company projected that its municipal system, estimated to cost about $660,000, would bring to Scranton 19 million gallons every 24 hours, delivered via a pipe 25 miles long. When Valley Water Supply brought its proposal to Scranton city councils in 1889, taxpayers balked at the high price.[24]

With drought redoubling fears about water, the D & H refused to let Valley Water Supply threaten its coal works near Carbondale, about seven miles south of Stillwater. The D & H objected when the water company, owned in part by independent coal operator John Jermyn, moved to seize five acres of D & H land. The D & H charged that the loss would drain it of hundreds of thousands of dollars, and almost as soon as the drought ended, the D & H began building its own dam at the site. When a judge granted the water company an injunction to stop construction, the D & H refused to end work. Noting that Valley Water Supply had let its charter lapse, the D & H argued that the water company had lost any right to take land. In response, Valley Water Supply filed for a five-year extension, but a judge denied the request in early 1896.[25]

A major employer in Carbondale, the D & H had a lot of support among residents. In 1889, several opposed Valley Water Supply enough to complain in the local paper that Scranton sought to usurp their water. Others said that water beyond the gap was rank with runoff from acid factories and drainage from Union Dale and Herrick Center. Besides, one noted, Roaring Brook already supplied Scranton. While Carbondale officials dithered over their own response in 1895, the D & H finished the dam. With the plug in place, locals thought that the levee might help prevent flooding and keep the river clean, particularly in the summer, "a condition that [had] not existed for many years." Others speculated that the lake would be "an advantage to Carbondale as well as a thing of beauty."[26]

A year after the dam went up, the newly formed Lackawanna Valley Water Supply Company took over its operation. With directors employed by the D & H, the company faced a local water system "taxed to its utmost." Planning pipes to valley towns, Lackawanna Valley Water periodically raised the dam, flooding more farmland, and drew water from the pond so often that it made for "a very unhealthy condition." Making matters worse, the directors dragged their feet in compensating the owners of submerged ground. By 1899, however, the new company had combined with Consolidated Water Supply Company, another business ruled by coal men.[27]

Lehigh Water Grab

New Yorkers drink from several watersheds. In the 1930s, crews completed an 85-mile tunnel to draw water to the city from the Delaware River and Rondout Creek. In the 1990s, residents of the Catskills rediscovered that in 1905, Albany had granted Gotham powers to control upstate pollution. In response, farmers and other landowners balked at new rules about manure spreading, paved surfaces, and buffer zones along streams and rivers. In 2020, the city and the U.S. Army Corps of Engineers visited northeastern Pennsylvania to study taking water from a reservoir on the Lehigh River. With droughts and sea level rise predicted, the releases would keep salt at bay in the lower Delaware, saving water in other reservoirs for New York.[28]

In 1895, water companies contended for the headwaters of the Lehigh. As Valley Water Supply faced off with the D & H on the Lackawanna, at least four companies staked claims to the Lehigh: Crystal Spring Water and Citizens' Water, of Wilkes-Barre; Lehigh Coal and Navigation; and Scranton Gas and Water. This conflict also involved dam building, land seizures, and court battles.[29]

With the drought deepening, Scranton Gas and Water (SGW) announced in October that it would implement long-standing designs to tap the Lehigh watershed, which extends into Lackawanna County. When SGW declared that it would build a dam near a proposed reservoir of Citizens' Water, the latter, which had "a posse there on guard," vowed that it would "take by force if necessary the Scranton company's selected reservoir site." Requiring pumps and a pipe to bring water into the Lackawanna basin, the SGW plan provoked Lehigh Coal and Navigation to join Lehigh Valley businesses and the Wilkes-Barre water companies to stop SGW from damming a Lehigh tributary. In its defense, SGW claimed that it would store water "only for use in time of severe drought," just when the others might need it too.[30]

The legal wrangling pointed to tensions between water for drinking and water for mining. A major anthracite operator, Lehigh Coal and Navigation claimed the entire Lehigh River for its canal; SGW argued that using water to serve drought-stricken people was more important than saving it for a ditch. With Scranton growing and the D & H and upstream towns needing tributaries of the Lackawanna, SGW asserted rights to any water it could get within Lackawanna County. When LCN responded that SGW had valid claims only in the Lackawanna watershed, SGW countered that it could take water from "any part of the state," including the part of Lackawanna County in the Lehigh watershed. In 1896, a court sided with SGW, which not long later pumped "700,000,000 gallons . . . from the Lehigh into Roaring Brook for use in Scranton."[31]

The 1895 drought hastened the consolidation of water companies. In early 1896, all but two providers between Scranton and Nanticoke merged into one company, Spring Brook Water Supply, forming a virtual monopoly in the Wyoming Valley. Scranton elites, coal operators, and railroad interests invested in the scheme, which had been planned for months. A spokesman explained that the company existed to offer "what everybody has been clamoring for, an abundant supply of good water the whole year round." Spring Brook intended one reservoir to supply residential customers and another to serve collieries and manufacturers. Unlike a proposed project of others to draw on the Susquehanna River, the new business took water from the surrounding mountains, claiming that "you cannot make anything out of the Susquehanna but a big sewer that receives all the waste and filth and offal from every town and village along its banks." Despite this description, a year later, Spring Brook built a pumphouse on the river to supply the Wyoming Valley.[32]

Droughts drive the thirsty to extremes, whether violent or extralegal, which may be why local water seizures echo larger grabs elsewhere. To sustain population growth, in 1913, Los Angeles snatched water from the Owens River, eventually sucking dry a valley filled with irrigated farms. After protracted battles to supply San Francisco, in 1923, California dammed Hetch Hetchy Valley, within Yosemite National Park. Celebrated as engineering marvels and points of national pride, big dams soon rose on the Colorado and Columbia Rivers: Hoover (1936), Bonneville (1937), and Grand Coulee (1941).[33]

These dams joined a global rush to raise high dams that began in the 1930s and crested in the 1970s. More than any other human endeavor, mega dams altered lands worldwide and became global "symbols of technological virtuosity and modernity." In 1960–1970, Egypt built the Aswan Dam; in 1982, Brazil and Paraguay dedicated the Itaipu Dam; and in the 1980s, India pushed plans for dams on the Narmada River. By 1990, opposition to Narmada dams had drawn international attention, prompting novelist Arundhati Roy to refute the "Local Pain for National Gain myth" in her essay "The Greater Common Good" (1999). Drowning all kinds of life, big dams not only narrow definitions of the common good but also excite debates about the effects of its expansion.[34]

Stillwater Runs Deep

As mining hit hard times, Stillwater became a flood control site. By 1941, efforts were underway to build an impoundment reservoir, but World War II intervened. No money could be spared for a dam in 1942 after a raging Lackawanna tore out rails, washed away coal piles, and ripped through businesses, idling the Doyle & Roth plant for days. When heavy rain hit the valley in 1947, people appealed to Congress for help, but the U.S. Army Corps of Engineers decided that a dam at Stillwater was not worth the price.[35]

In 1955, Hurricane Diane changed minds. Soon scrambling for dam sites within the valley, army engineers discovered that mining had disrupted the geology too much. In 1957, the federal government seized about 100 acres near Union Dale, and the next year, construction began on an earthen levee about a mile upstream from the D & H dam. Nine months later, while crews raised the dam, the Susquehanna River poured into the Knox workings.[36]

Today, Stillwater Dam ties together water-energy grids. Carrying Marcellus shale methane south, a pipeline buried in a former railbed supplies a billion-dollar power plant in the Lackawanna Valley. At Stillwater, this line intersects the Tennessee Pipeline, which crews laid in 1955; striking east beneath the lake, the Tennessee feeds gas to New York and New England. Crossing under rail routes that hauled anthracite, the pipe carries not only gas from Marcellus shale but also methane from refineries in Louisiana and Texas.[37]

Collecting the headwaters of the Lackawanna, the reservoir at Stillwater appears placid, apparently empty of significance apart from capturing and regulating flows. Its meaning beyond its work, however, becomes visible when one sees it as a spot of time, a swirl of events that touch others, near and far. A spot of time is deeper and more extensive than a series of moments in place. To understand it requires embracing the form that it assumes, whose expression must always be in some sense fluid because interwoven times and places, all at once, everywhere, constitute "reality." To write a history of a spot of time is to acknowledge ambiguity, to know form as content, and to search beneath surfaces. Paying attention this way, writing this way, is "literary."

Narratives of climate history might best be fluid because rigid forms may soon crack under the weight of global warming, whose scale and effects make it too ubiquitous to see, too massive to capture, and too harmful to ignore. More easily than most, flexible forms can follow its flows. Using them, historians can trace how we got here, teaching us not only what happened, how it happened, and to whom and why but also the thinking that gave rise to it all.

8

Working Water

Water shapes the earth. It evaporates, freezes, and flows away, and it has the power to rust, rot, and erode. Hampering and enhancing our efforts to survive and thrive, it takes form as resource, laborer, and product, making it sometimes a focus of contention between workers and business owners. Just as water is at once nature and culture, so it is nature and economy.

In the northern anthracite coalfield, the work of water sparked conflict, some of it bloody, especially during strikes in 1900 and 1902 when miners and managers battled at washeries. Meanwhile, as cities planned municipal services, private water companies—financed with money made in mining—fought for supremacy in the Lackawanna watershed. With coal production and water supply networks expanding, pipes and ponds became numerous—and litigious—features of a torn landscape.

Remining Waste

Prior to 1880, collieries tossed aside small sizes of anthracite, which accumulated as piles of waste coal, or culm. Blossoming beside breakers as reminders that not all anthracite was marketable, banks of unsellable coal could amount to as much as 70 percent of a colliery's output. In the early 1880s, Lackawanna Iron and Coal Company discovered how to use culm to generate steam, and by 1889, the Scranton Board of Trade advertised small-sized anthracite to East Coast and Midwest manufacturers. Answering the demand, coal operators struck gold in waste banks; new marketable sizes made up as much as half of some piles. One dump in St. Clair was 75 percent usable.[1]

The key to turning waste into cash: washeries. Making their first appearance in the Anthracite Region in 1891, they used water to mine culm banks and separate rock from coal. Often portable and sometimes converted breakers, they required 500 gallons of water, about 2 tons, to produce 1 ton of coal. With an average output of 400 tons per day, the typical washery consumed 200,000 gallons daily and ate about 10 percent of the fuel it recovered. By 1899, 20 operated in the northern field, widening revenue streams drained by rising costs of deep mining. Although the average washery employed 29 men and cost 30 cents per ton to run, owners soon found that their survival depended on selling smaller sizes.[2]

Washeries were akin to breakers. At waste banks, workers wielded water hoses to loosen culm, which coursed to sluices that carried the slurry to a "scraper line" that elevated the mess to the top of the washery. From there, it dropped through sizing screens into a reservoir. As an engine roiled the water, coal rose to the top and flowed one way; rock sank and flowed in another. Slate and coal dust left the building borne on flows of water, or slush, adding another stream of pollution to colliery output. Some companies flushed this fine silt directly into abandoned mines via boreholes, a disposal system that itself took a lot of water. Collieries piled the rock waste, and some washed silt into creeks, raising their beds. No matter the way, water did the work before disappearing.[3]

Workers watched warily as washeries went up. After all, every ton of washery coal decreased demand for fresh-mined anthracite. In 1892, companies shipped 90,495 tons of washery coal; in 1901, they moved 2,567,335 tons, a 27-fold increase, before climbing to 3,846,501 in 1906. Although reliant on deep miners, washeries employed largely unskilled labor. In 1900, washery workers earned $1.30–$1.49 per day; depending on the calculation, mine laborers made $2.00 to $2.25 and miners $2.25–$2.50. Eating into waste banks, washeries also took coal from poor families who picked culm to heat homes and cook food. Adding to worker discomfort: deep miners worked semi-autonomously while washery men fell under close management supervision.[4]

Washeries nettled union members during the 1900 strike, September 17–October 28. They wanted washeries stopped because they were "turning out considerable coal" and shipping "many carloads." In parades, strikers hoisted banners that read, "Do not handle washery coal, that is what the company stole from the miners." Although labor organizers initially permitted the Grassy Island washery in Olyphant to run, they called for it to close after discovering that the company was shipping coal from there. In late September, workers in Wilkes-Barre stoned nonunion men at a washery and tried to derail a trainload of wash coal. On October 22, the first strike-related riots in Wilkes-Barre erupted at the Empire washery: 2,000 to 3,000 workers seized the operation, drove off the Coal and Iron Police, and delivered anthracite to local widows.[5]

Both sides rallied around washeries. When anti-strikers observed that washeries could supply "culm-burning locomotives," union men demanded that those remining stop, declaring that it would be "a suicidal policy if they permitted the washeries to work." Some mine owners asserted that washeries would run "if they had to call for all the state troops to protect them." Although Pennsylvania Coal Company closed its mines during the 1900 strike, it kept two washeries open "to furnish coal for pumping purposes." For the same reason, the D. L. W. put mine bosses in Scranton to work to keep three washeries running. Union officials opposed their operation, knowing "the great loss that would result from a stoppage of the mine pumps." In addition to hamstringing collieries, however, closing washeries threatened work at nearby trolley, heat, and power companies, a potential public relations problem for both sides.[6]

Scranton Gas and Water

In 1890, most U.S. towns of more than 30,000 boasted public waterworks; by 1897, over 80 percent of major American cities had municipal systems. In the Lackawanna Valley, though, private concerns ruled. As mining expanded, populations increased, and demand grew, coal barons maneuvered to take control. Accelerated by the 1895 drought, this chess match featured two millionaires: William Walker Scranton, who capitalized on inherited privileges, and John Jermyn, an immigrant laborer who amassed a fortune from scratch.[7]

The Scranton family long stood at the heart of the local water-energy nexus. In the mid-and late nineteenth century, Scranton companies mined coal, made steel, and distributed water. Brothers George and Selden built iron furnaces in the early 1840s, founded Lackawanna Iron and Coal Company in 1853, and established Scranton Coal Company in 1854. Their cousin Joseph H., a partner in SGW, led Lackawanna Iron and Coal from 1858 until 1872. His son William Walker (W. W.) unfortunately quarreled with company officers and could not succeed his father as president. Instead, in 1881, he started the rival Scranton Steel Company, which Lackawanna Iron and Coal acquired in 1891. After the sale, W. W. turned his attention to SGW, expanding it into a regional monopoly by 1905. His son, Worthington, sold the business in 1928 to Federal Water Service Corporation of New York, yet another affirmation of the valley's colonial status in relation to the port city.[8]

A "rugged type of man" with "giant strength," W. W. Scranton (1844–1916) had a reputation in business as "aggressive, setting his mark and never failing to achieve it." His wealthy father made sure that he was well educated; an alumnus of Phillips Academy, the boy graduated from Yale (1861–1865) before coming home to work as an assistant at the iron mill. After two years learning the ropes, he began managing the plant, earning a name for refusing to back down from a fight. Workers did not know him as a friend of labor.[9]

In the strike of 1870–1871, W. W. accompanied a squad of National Guardsmen who were escorting workers at the Briggs colliery home when they confronted a crowd of about 400 strike sympathizers. After someone threw a stone, a soldier fired, killing two, and more may have died had not W. W. "knocked up the rifles and cried 'Don't fire boys, that will do.'" Although he was arrested, Scranton was acquitted of any role in the deaths.[10]

The national railroad strike of 1877 also spilled blood in Scranton. On August 1, a crowd of mine workers meeting at the Sauquoit mill grew angry as someone read aloud a letter from W. W. Scranton. Likely a forgery, the note dismissed the men as slaves and threatened a steep wage reduction. Incensed, the gathering attacked work sites before spilling into downtown, where a citizens' corps met them with Remingtons. At the front marched W. W. Veterans of the Civil War, the corps fired three volleys, killing 3 and wounding 25 or more. The next day, a National Guard unit patrolled the city.[11]

The Scranton family's iron and water interests undermined as much as helped one another. Organized in 1854, SGW supplied the village, railroad shops, and ironworks from Roaring Brook and the Lackawanna River. By 1866, mine acid made the water undrinkable, so the company began a retreat into the hills, building its first dam near the iron mills. By 1870, acid had poisoned this pond so much that rust-eaten boilers at the furnaces exploded, killing 11. To escape the pollution, in 1873, SGW located its next lake beyond the measures.[12]

SGW regularly built reservoirs. With the completion of Elmhurst Dam in 1889, the company dammed Stafford Meadow Brook in 1893 to create Williams Bridge Reservoir. In 1894, SGW planned a pond at Burnt Bridge that would hold 1.4 billion gallons, ensuring that the company could boast a water supply to last more than 200 rainless days. Completed in 1898, Burnt Bridge has been naturalized as Lake Scranton. Expanding as the city grew, by 1902, the privately owned SGW served Scranton from 10 reservoirs, which contribute to the "excellent supply of good water" most valley residents enjoy today.[13]

Contending for Washeries

During the 1902 strike (May 12–October 21), owners and workers fought over washeries because they were key to controlling collieries. Operators wanted mines dry, which meant running washeries to fuel steam engines that powered water pumps. Surface sites, washeries could be kept in operation because they did not depend on skilled workers to cut coal, raise cars, or make repairs, so when pump men walked, managers hired scabs and pressed clerks, foremen, and fire bosses to man pumps. Early in the strike, companies cooperated to open several washeries and two collieries for their own use; to cut the number of men needed to operate drills, they relied on mines powered by electricity. Despite their efforts, bosses, clerks, and nonunion men could not lift all water in all mines; some filled, and a few were abandoned. Adding to anxieties about flooded mines, rain fell periodically, once for a solid week, September 24 through October 1, dropping six inches across the region.[14]

Although pump men, firemen, and engineers usually stayed at work during strikes, the union convinced most to walk on June 2. The threat of flooded mines, organizers believed, would bring owners to the table. To stop all pumps, strikers repeatedly marched on washeries and harassed the men who ran them. The union wanted washeries shut because their operation meant that strikers could not halt coal production; particularly galling, the Butler washery, open despite attempts to close it, broke production records.[15]

Owners and union officials knew that the decision of the pump men to work or walk would be a key moment, which may be why both sides claimed victory on June 2. Depending on whether owners or union leaders reported the numbers, anywhere between 45 and 75 percent of pump men stopped working. Strikers had silenced pumps, gained important allies, and shown that their grievances were widespread. Owners had kept pumps operating, saved mines from flooding, and shown that they could run collieries without strikers. From then on, as the *Scranton Republican* noted, the unrest entered a "new phase . . . the era of active hostilities." After the June 2 walkout, the union convened committees at each colliery to keep engines and water pumps idle.[16]

Some pump men chose not to strike. On May 24, those at the D. L. W. Storrs Nos. 1, 2, and 3 voted to stay on the job. At Erie mines in Forest City, crews remained at work. On June 24, a dozen engineers and pump men in Nanticoke returned to their posts because Susquehanna Coal Company "had a desperate time preventing floods," despite importing nonunion workers. Mine owners declared their return a victory.[17]

Caught in the crossfire, other industries fell silent in the war over washeries. Union organizers initially allowed a few to run to meet local needs but soon moved to shutter them. For the first months of the strike, a dozen union members worked three eight-hour shifts at a washery just north of Carbondale. After strike officials ordered the men off the job in July, shipments of coal stopped to the nearby silk mill, which daily needed a 15-ton supply. Lack of fuel also closed an electric plant, a brewery, a creamery, and laundries, putting several hundred men and women out of work. Loss of the electric plant in turn affected operations at Carbondale Milling Company and a bobbin works. When the Sauquoit mill in Scranton stopped, over 1,500 were idled.[18]

Attention riveted on washeries for several reasons: they supplied fuel to pump mine water, running them made money for operators, and independent contractors often leased them, which usually meant that they could little afford to close them. Although mine owners claimed that they worked washeries only to supply their own needs, mainly to run pumps, they sold coal from them when they could. In the conflict to control pumps, which meant controlling water, workers and managers brandished weapons of war: rifles, barbed wire, stockades, and searchlights.[19]

Consolidated Water Supply

Stoked by coal mining, breathless growth hindered Scranton in meeting public utility needs. Between 1840 and 1914, the city's population jumped from a handful to 135,000, with the number of collieries in town leaping from none to 27, or about 1.3 per square mile. In 1904, Scranton boasted 116,000 residents, five steam railroads, 30 mines, and "one hundred forty manufacturing interests." As water demands ballooned, W. W. Scranton confronted coal barons who financed SGW competitors.[20]

Coal operator John Jermyn challenged W. W. Facing "limited opportunity for securing an education" in his native England, Jermyn came to the United States at age 22, landing his first job with the Scrantons. Unlike W. W., whose father ran Lackawanna Iron and Coal, Jermyn rose from laborer to millionaire with apparently little more than hard work and good luck. Although W. W. may have had rocky relations with workers, Jermyn maintained good connections to former employees and was often observed exchanging memories with "a 'butty' of ye olden days."[21]

Unlike many other early mine owners, Jermyn refused to sell his colliery to a railroad; later, like the Scrantons, he straddled the water-energy nexus in expanding from coal mining into water mining. Prophetically, one of his first jobs was to drill a water well, and he was a major partner in Valley Water Supply. After he died on May 29, 1902, his son Joseph succeeded him as head of the family mine holdings, Jermyn & Company, adding to his duties as president of Consolidated Water Supply Company.[22]

In 1897, John Jermyn incorporated Consolidated Water, which the state chartered in 1899 to supply residential, commercial, and manufacturing customers. Headquartered in Scranton, the company absorbed 10 small water outfits, among them Lackawanna Valley Water Supply, which gave it control of Stillwater Dam, and Crystal Lake Water Company, which supplied much of Carbondale. Dominating water systems at the upper end of the Lackawanna Valley, Consolidated Water not only commanded the watershed from Archbald to Ararat but also planned to send water from Stillwater to Scranton.[23]

Financing the feat proved impossible. Consolidated Water first ran into trouble in Carbondale because of rate hikes. In May 1899, city officials decried increased hydrant fees, and by August, residents howled about a near doubling of household bills. In response, the town explored municipal ownership, a court appeal, and the creation of a rival company.[24]

Public water went nowhere in Carbondale, despite residents voting overwhelmingly in 1900 (1,670 to 328) to issue bonds to build a water plant. Consolidated Water, town taxpayers, and the D & H filed suit to stop it, and when the city planned to draw on the watershed of the Jermyn Water Company, John Jermyn said no. In 1902, a judge upheld an injunction to keep Carbondale from issuing bonds; a year later, the select council voted against appealing the ruling.[25]

When Consolidated Water raised rates, locals did more than fight back at the ballot box; they found new sources. Several established Carbondale Artesian Water Company, and individuals with enough money sank artesian wells. A screen maker for collieries, E. E. Hendrick drilled 500-foot wells at his business and at his home. Klots Silk Mill drew 75 gallons a minute from a 600-foot well, but a pump tripled the flow. Not far from Carbondale, two artesian wells supplied D & H No. 3 washery, which needed 300 gallons per minute. In 1901, Carbondale interests incorporated the Belmont Water Company and built Belmont Reservoir, a three-acre pond in Fell Township whose name likely alludes to the Meredith estate.[26]

John Jermyn died at a turning point in the 1902 strike. On May 27, his son Joseph asked the pump men at the family collieries in Jermyn whether they would strike on June 2. When no one replied, he fired the lot. On the Thursday that Jermyn passed, May 29, union officials anticipated that most pump men would walk, but owners believed otherwise after a meeting of pump men from several collieries, including the Jermyn mines, voted to stay on the job. On the Monday morning after Jermyn's burial, however, many at the pumps joined the strike. At the Saturday funeral, W. W. served as an honorary pallbearer.[27]

Washeries Under Fire

In the 1902 strike, washeries served as major flashpoints of unrest, which manifested as mob protests, gun battles, and burnings. The first strike violence happened on May 20 when union workers drove off scabs running the Butler washery in Pittston. The next day, strikers exchanged gunfire with nonunion men, and owners threatened more strife if strikers called out pump men to fill mines. Meanwhile, east of Pittston, in Avoca, shots rang out at the Elmwood washery; in Olyphant, 1,500 workers forced the closure of the Grassy Island washery.[28]

Strikers fought operators at water pumping stations. To stop the National washery, men dynamited its riverside pump. Despite the damage, the washery kept working with water drawn from a nearby mine; when men at the mine stopped working, the manager tapped storage reservoirs. On August 11 at the Pancoast washery in Throop, strikers stormed pumps, exchanging shots with deputies. Officials declared that the washery would remain in operation even "if the militia must be summoned." Responding to the incident, the *Scranton Republican*, generally sympathetic to mine owners, chided that "one riot will do more to make the strike a failure than would a dozen washeries running at full blast."[29]

Suspicious fires destroyed washeries. When the Bellevue, the largest in the Anthracite Region, burned in early August, company officials suspected arson. Nonunion men employed by Scranton Coal Company started a fire that destroyed the three-year-old Briggs washery in Keyser Valley, which daily produced 1,000 tons of coal. The next day, the Capouse washery burned, despite "coal and iron police watching the colliery and a couple score of men engaged about the place." Not 10 days later, another "mysterious fire" destroyed a Hazleton washery after the owners attempted twice to reopen it. About three weeks after this, a washery in North Scranton burned.[30]

Owners collected scab labor and made washeries into "armed camps." Early on, at least seven employed "imports" who were "housed and fed" on-site. The Butler washery hired "foreigners gathered from the slums of New York," and Erie Coal Company shipped 100 men from Philadelphia to Pittston "under the pretense of employing them in the construction of the 'Cannon Ball' railroad." Deputies hustled the new arrivals into the walled No. 6 washery, but soon after the gates locked, the men realized the ruse, refused to work, and "scrambled over the fence, so anxious were they to get away."[31]

Turmoil plagued washeries in other ways. In Duryea in August, 50 laborers, a barbed-wire barricade, and Coal and Iron policemen could not keep the Warnke washery open for more than a day. As a crowd marched on the building, guards fired, wounding one or more strikers. Some attributed the August 19 dynamiting of a home in Pittston to the fact that boys in the family worked at No. 6 washery. To keep a washery near Wilkes-Barre from reopening on September 19, someone blew the dam that supplied it with water. That Lehigh Coal and Navigation kept two washeries in Panther Valley at work led to the shooting of a strike leader; uproar over his death prompted the local sheriff to seek help from the governor, who dispatched a National Guard regiment. Some owners—including at Pancoast and Johnson—installed searchlights to guard collieries. In mid-July, the D. L. W. advised residents of Taylor to leave their homes because washery wastewater would soon flood the Flats section.[32]

Despite protests and violence, companies in the Lackawanna Valley kept seven washeries at work in late June, more than were in operation at any time since the strike began. Adding to tensions, the D & H opened a new one mid-month, and by the end of August, owners had 11 working. With violence at them escalating, on September 23, National Guard units set up camp in the valley.[33]

Good Manners and the "King o' New York"

The humorous short story "Good Manners and the Water Company" suggests how everyday people might respond to arrogant water monopolies. Set in the Wyoming Valley and appearing in *Century Magazine* in 1908, the story was authored by West Pittston resident Emily Johnson, who knew something about ethics: active in aid to poor immigrant families and for 35 years the secretary for a Luzerne County judge, she saw corruption in the coal industry up close.[34]

As its title implies, the story is about behavioral norms: there are good and bad ways of doing business. Valuing informal codes of human interaction, the narrative criticizes impersonal rules that the water company wields as a club. Large enterprises succeed, the story suggests, by integrating themselves into the community, an assertion that touches the coal cartel as much as water trusts. Underscoring this ethic, Mrs. Kinshalla's name alludes to agreed-upon family behaviors, "kin shall a." The rigid agent she confronts goes unnamed.

Due to the promotion of a long-serving bill collector, a stranger shows up in Duck Hollow to knock at Mrs. Kinshalla's door. When her daughter, Irene, answers, she tells the "new Scotch lad" that she "never buy[s] a thing off agents. . . . You always get cheated if you do" (57). Unfamiliar with the Irish neighborhood, the novice collector refuses to banter because "his temper was bad" (57). His lack of engagement leaves standing Irene's insinuation that his employer cheats.

The new bill collector represents a type of company: humorless, overbearing, solely profit driven. When Irene asks him whether he's "King o' New York," he replies, "I'm the watter comp'ny" (58), lines that nod to monopoly control of local anthracite and water. Angry that Irene won't pay, the agent tells her that unless he gets the money, he will "toorn yir watter off accordin' to steepulations pervided in the contrac'" (58). Irene claims that the bill was paid and dares him to deny them access. To teach her a lesson, he shuts off the water, noting that "expeerience is the fine teacher" (58), a remark that boomerangs on him later.[35]

Threatening community bonds, the Scot tells Mrs. Kinshalla that if her neighbors provide her with even a cup of water, their supply will be cut off too (59). After Kinshalla helps herself at the sink of Mrs. Loughney, the Lougheys and their renters lose access (60, 61). Families soon fight over the severed services, "hat[ing] one another singly as much as, collectively, they hated the water company" (61). Despite civil strife and company blacklisting, neighbors distribute water to each other, "in obedience to their prayer-books" (61), codes at odds with corporate rules.

Good Manners and Scranton Gas and Water

At the turn of the century, spiking water prices sparked industry pushback. In late 1899, the Scranton Board of Trade requested that SGW settle on a "uniform and lower meter rate to manufacturers." High water rates, the board reasoned, kept businesses from locating in the city, and current ones complained that "water for steam purposes cost about as much as fuel in certain establishments," leading a few firms to cut rates by deliberately wasting water. The board wondered, "If the water rates here are comparatively low, why does the Lackawanna Iron and Steel company pump filthy water from the Lackawanna river for use at their mills, and why are so many concerns putting down wells and using water thus secured wherever possible?"[36]

SGW responded with a new sliding scale that charged large consumers more, but they refused to set a lower uniform rate. Disappointed, the board observed that SGW "had not an exclusive franchise, [and] any other person could form a water company." At least one member urged that "an attempt should be made to secure another meter company's entrance into the city." The result: in 1899, collieries and manufacturers who accused SGW of price gouging created Meadow Brook Water Company, which SGW promptly purchased. Threatened by the Scranton Board of Trade and keeping a close eye on John Jermyn's Consolidated Water, W. W. fought yet another competitor, City Water Company.[37]

Just as Lehigh Coal and Navigation Company waved its charter in 1896 to keep SGW from drawing on the Lehigh watershed, SGW dusted off its own franchise in 1903 to stop City Water Company from supplying Scranton. Confirming expectations that W. W. would mount a "stubborn resistance," SGW went straight to court to kill the upstart on the grounds that only SGW had the rights to supply the city. At issue: whether the SGW charter made these rights exclusive, a claim City Water denied, responding that competition would lower rates. Accepting the argument that Scranton had room for two water companies, the state granted City Water status in September 1903.[38]

W. W. had a history of friction with a founder of City Water, former mayor William L. Connell. Connell's uncle, also William L. Connell, a coal operator, had entered politics a dozen years before. After the *Scranton Republican*, run by W. W.'s brother, Joseph A., criticized the Connells, the coal baron and a few friends founded the *Scranton Tribune* in 1891, and Connell used it as a friendly voice in his successful run for Congress in 1896. In December 1901, as arguments for municipal water roiled Carbondale, then Scranton mayor Connell invited residents to write to him with their thoughts about the advisability of a public water system; most of those who submitted letters favored the idea. Establishing municipal water amounted to contracting with a new company, buying SGW, or building a city-owned system. W. W. was likely livid.[39]

Complicating his life more, City Water and Consolidated Water cooperated to challenge him. To meet the city's demand for 20 million gallons per day, City Water had an option to draw on the daily supply of 75 million gallons at Consolidated Water. The combined companies toyed with the idea of supplementing their reserves from lakes and new wells near the headwaters of the Lackawanna. They also planned a reservoir at Stillwater, another at Brownell swamp, and pipelines to Scranton.[40]

The battle to stop the SGW monopoly ended with its expansion. When City Water could not act on its option with Consolidated Water, the Jermyns entered negotiations with W. W. Their June 1905 deal handed SGW control of much of the watershed, from the headwaters at Ararat to Scranton, amounting to 14 reservoirs and about 100 square miles.[41]

Wider monopoly power did not soften W. W. In 1907, the Scranton Board of Trade again aired anger at him: "he appears unmindful of the fact that the business of his corporation is effected with the public use, and, to that extent, subject to the visitation and control of the people." In 1902, after the city placed a licensing tax on corporations, SGW raised its annual household water rate from six to eight dollars. When the city repealed the tax in 1907, SGW returned the charge to six dollars, as promised, but a few months later upped industry rates. To complainers, W. W. replied that he had kept his word, but he pointed out that he never said that he "would not endeavor to make up for the loss to some extent by the change of rates in other directions." As for municipal ownership, he simply dismissed the idea, noting as evidence the poor condition of city bridges.[42]

Good Manners and Johnny Selden

"Good Manners and the Water Company" encourages guerilla tactics to fight water monopolies. Echoing labor organizers, Mrs. Kinshalla exhorts Loughney to stand up for herself against the "dhirty Trust, that there water comp'ny. An' the trusts has had their day; the people are about done with 'em. John Mitchell said so himself" (60). The mention of Mitchell points to unionization, in particular to the 1902 anthracite strike, which he led. Rather than organizing into a formal body to demand redress, however, Duck Hollow residents deal with the water famine by hauling some at night.

Residents do try aboveboard means. Irene's husband, J. Addison Kohlmesser (58), a name that associates the couple with anthracite, defends his wife against the water trust by writing to a newspaper friend in New York (61). Kohlmesser's letter, which responds to "corporate insult," is published on page four (61), suggesting that his complaint is not breaking news. No one from afar, however, comes to the community's aid.[43]

To the rescue: the local volunteer fire department, a collection of clerks and coal miners. As Mrs. Loughney washes clothes, her stove cracks, forcing her to dump hot coals outside on an ash pile (61), which soon ignites her woodshed and chicken coop (63). Mrs. Kinshalla says that she'd put out the fire herself, "only for what 'ud happen if the water comp'ny was to see us a-stealin' water . . . in broad daylight" (63). After the firemen douse the blaze, Irene invites them inside (64). As the men crowd the kitchen, Mrs. Kinshalla points to "clo'es washed an' not rinsed, an' there's the tub empty an' waitin'. Not the drop o' clean cold water can we get till after dark" (64). The sympathetic firemen fill the washtub (64).

Irene proposes that the volunteers "be called down here every Monday. . . . Wash day is the very time a fire'd be most help again' the water comp'ny" (65). In exchange for a midday meal, the men agree to answer a fake fire call every week (65). The informal deal satisfies everyone, and soon "the work of mercy proceeded through the waterless neighborhoods" (65). The plan hoses a company that Kinshalla repeatedly calls dirty. Linking coal, clothes, and water, she foresees the next few Monday "fires": one a chimney blaze, the other in a coal bin (65).

The old and new agents represent two faces of the company. Outwardly, the old agent, Johnny Selden, is the company as it was under the "old management" (65), a member of the community. Gregarious, he shares stories, flirts with Irene, and knows details of the family's history. Realizing that the Scotsman—the new management—needs a lesson in appearances (65), Selden may "get him moved out on the reservoir gang awhile. . . . He deserves it for his manners" (66). To end the "water-fight" and bring service back to the neighborhood, Selden will return the next day with the Scot, who might be publicly humbled to reform "his complainin's an' spitefulness" (66). Unlike the Scot, Selden understands that people are more apt to follow the rules when they like the local agent, the public face of the company.

"Good Manners and the Water Company" reflects early twentieth-century anxieties produced by increasingly powerful and imperious corporations. Although the story ends with the water company making peace with the community, systems of corporate control remain undisturbed, if not empowered. Johnny Selden is only a friendlier face of the water trust; his promotion marks a move by the company to make itself "friendly," which the story reveals to be more profitable than the Scot's myopic, all-too-visible enforcement of rules.

Mining the water commons, the coal monopoly created a loop that poisoned the commons. A few days after the 1900 anthracite strike ended, the Scranton street commissioner closed the Mount Pleasant washery until the owner, Fuller Coal Company, stopped dumping culm into fields near Love Road, where a stream carried the coal to the Lackawanna River. As the 1902 strike dragged on, a judge okayed Elk Hill Coal and Iron Company to run washery waste to the river via a ditch that Archbald officials claimed overflowed. In 1913, the Gravity Slope workings siphoned water from a dam in the Lackawanna and pumped the resulting slush to a waste bank, but water still carried culm to the river, where it accumulated, despite the swift current. Piping mine water to its washery, the National mine sent slush underground but directed wastewater to the river in "a large stream which is dirty and full of culm."[44]

In a double loop, mining culm dumps with washeries cut costs so that companies could afford to mine further and deeper, which raised pumping costs, which meant mining more waste, which further polluted waterways. Mining culm helped keep the industry afloat until it developed improved cleaning technologies, which kept it alive until the cost of moving coal intersected the cost of moving water. As happened here, the world may be fast approaching a time when the cost of energy meets the cost of water. Living then with the damage will require patience, forbearance, and empathy, qualities too often in short supply.

9

Transforming Water

I was six, maybe seven. After herding cows out into the pasture behind the barn, I struggled to close the gate. No matter how I gripped the rubber end, the single strand of wire would not meet the barbwire loop. Frustrated, unthinking, I snatched the ring; screaming, I couldn't let go. Appearing at the barn door, my father tossed his shovel and stepped to the milk house. I fell. My left leg, it wouldn't stop twitching.

"You're all right," he called.

I stood, shaking, and stepped toward him.

"Close that gate," he said. "And don't forget to plug it in."

Hesitating, I pulled wire and loop together, made my way to the milk house to recharge the fence, and limped into the barn, my leg quivering, my mind processing a lesson in power.

A wall outlet in the milk house charged the wire, which wouldn't have happened until the late 1940s, when the first lines reached the farm. Before then, my grandfather lit house and barns with batteries. Wires and insulators from those days clung to walls and beams, and I found, now and then, other odds and ends, one a small wood-cased battery.

After that day, I kept clear of live wires. I've been shocked since, but nothing has come close to the fright of being bound to that loop, which, I've learned, tied me to a grid powered, at least in part, by the working waters of Lake Wallenpaupack.

Demographics may soon knot entanglements of water, electricity, and coal. Since 1950, a growing human population and an expanding urban footprint have ballooned water and energy demands. By 2050, three billion more people may use another 4.5 billion acre-feet, which will spike electricity generation, which will increase fossil fuel burning. Accommodating these needs may not be possible, though, given that offering everyone the opportunity for a decent life runs up against a rapidly diminishing natural world. Endangering human and nonhuman systems that rely on some stability to function, this hard reality may soon visit on us what we have inflicted on the rest of life: since 1970, wildlife numbers have plummeted by more than two-thirds. Catastrophe surrounds us.[1]

Lighting human lives darkens earth systems because, unfortunately, electricity does not generate spontaneously. Although hydro and nuclear contribute, fossil fuels make 90 percent of juice in the U.S., mainly in water-cooled power plants. Just as unfortunately, the grid distributes not only electricity but also pollution. In 2011, power generation coughed up more than a third of the human-made carbon dioxide released in the U.S. In 2013, steam electric plants dumped into American waters more than half of all toxins produced in regulated U.S. industries.[2]

Electricity dominates modern power and light. Transmission lines stretch across landscapes, wires snake in and out of places large and small, and even something as abstract as "the cloud" relies on currents streaming across miles of copper. The ease of flipping a switch or pressing a button suggests how hard it might be to let go of fossil fuels, maybe especially here, in northeastern Pennsylvania. Grasping the history of local power generation lights up tight ties between coal, electricity, and water.

When the 1902 strike unveiled the power of pump men, mine owners turned to electricity. Operators sought it to drill and cut coal, power cars and hoists, and run crushers and water pumps. Right after the labor unrest ended, Erie Coal Company designed electric collieries to cut breaker boys from payrolls. By 1910, most hard coal outfits used electricity; four years later, industry transmission lines extended 169 miles. Running currents between operations, mining companies generated their own power, which in the 1920s consumed 7 to 8 percent of total coal production. Just in case heavy rains increased pumping needs, however, they also kept plugged in to public sources. Largely self-contained, colliery powerhouses burned the smallest sizes of anthracite.[3]

The D & H may have targeted pump men. As the strike entered September, Philadelphia entrepreneurs proposed a dam on Wallenpaupack Creek to generate power for D & H mines in Wilkes-Barre, about 46 miles away. Using electricity would make pump men unnecessary because the powerhouse in Hawley would need no engineers in the Wyoming Valley. With energy to move mine water originating at the dam, not the mine, "coal strikes [would] be entirely eliminated." As a bonus, hydropower, or white coal, would annually save the D & H "the expense of purchasing 1,000,000 tons of coal."[4]

Although the Philadelphia plan stalled, others jump-started it. In 1910, just as anthracite neared peak production, water magnate Louis Watres decided to bring electricity to the Lackawanna and Wyoming Valleys. To collect runoff from a 200-square-mile watershed, he foresaw a 40-foot-high dam on Wallenpaupack Creek that could corral 40 billion gallons. Designed to generate 10,000 horsepower daily, the project was backed by, among others, coal baron Clarence Simpson.[5]

Watres had fought labor battles before. During the riot of 1877, he served as a private in the Scranton City Guard, likely placing him in the melee with W. W. Scranton. As the 1900 strike unfolded, he played a major role in building the 109th Regimental Armory in the city, and in the 1902 troubles, he commanded the 13th Regiment of the National Guard when it deployed to the Lackawanna Valley. Camped at Olyphant, not far from the Grassy Island washery, the unit patrolled the upper valley, helping to reopen a handful of mines by mid-October. As Watres kept an eye on strikers, crews completed a mansion for him on East Mountain that offered wide views of the city.[6]

Watres worked in coal, politics, and water. In the mid-1870s, he co-owned Green Ridge Coal Company, but he soon pivoted to politics, serving as state senator and lieutenant governor. A founder of Valley Water Supply, he was also a partner in Spring Brook Water Supply Company, created in the wake of the 1895 drought. In 1916, the extent of his water interests, concentrated in Luzerne County, rivaled those of W. W. Scranton. In 1928, Federal Water Service Corporation took control of the two businesses, which operated as Scranton-Spring Brook Water Service Company.[7]

The Watres scheme also failed, but others kept it alive. Planning to electrify operations, the D. L. W. railroad negotiated with Watres in 1922 to buy the Wallenpaupack site. After talks collapsed, Pennsylvania Power & Light Company (PPL) stepped in to purchase the land in 1923. Formed in 1920, PPL catered first to cement, steel, and anthracite mining companies and during World War II was "the largest single user of anthracite coal in the world." Turning to bituminous in the 1950s, PPL today supplies power to 10 million in the U.S. and the United Kingdom and delivers natural gas to customers in Kentucky.[8]

PPL raised the dam (1924–1926) at Wallenpaupack during a period of national debate about whether power generation should be a private or public enterprise. As water filled the new reservoir, observers linked the lake to Pennsylvania governor Gifford Pinchot's Giant Power plan to generate electricity at bituminous mines and hydropower sites. Seizing on the idea, third-party Progressives "made public control of hydropower and coal a defining feature of [their] national agenda." Pinchot's "first concern," however, was to bring electricity to the farmer. Private power companies launched a successful propaganda campaign to kill the plan.[9]

Heat Exchange

Engineering: "We consider ourselves as much of an engineering firm as a manufacturer of quality heat transfer equipment."

—Rohit Patel, vice president, Doyle & Roth

Located in Simpson, just north of Carbondale, Doyle & Roth sits on land that the D & H bought in 1825, two weeks after crews began building the canal. My brother Bob spent most of his working life there as a pipe fitter. He hated it. He loved the farm.[10]

Workers at the plant make heat exchangers. Obeying the laws of thermodynamics, an exchanger transfers heat between fluids that are not in contact. The company mainly builds recuperative indirect exchangers in the shell-and-tube type. In a recuperator, fluids flow through their own channels within the exchanger; indirect transfer means that heat conductors, such as steel plates and pipes, separate the fluids. In many Doyle & Roth exchangers, heat transfer makes water into steam, a two-phase change common in boilers and condensers. Although the plant's time clocks, benders, and hole punchers used the most electricity, only arc welders and handheld grinders made it visible.[11]

The summer after I finished high school, I worked at the plant. I did odd jobs: fetching tools, sweeping floors, moving bundles of pipe. Usually, though, I ground metal. When one 10-foot heat exchanger didn't quite accommodate a collection of pipes, I sledgehammered the outer shell and ground the inside, which meant sparks shooting up the arc until they reached the top, only to fall on my exposed neck. Every so often, the foreman would stop me, climb inside, and rotate a steel stick. After scores of tries, the rod finally turned freely. The work paid just above minimum wage.

At my first off-the-farm job, I noticed how rigidly clocks divided the day. My brother and I left home at 6:30 every morning so that we would punch in right at 7, just as the morning siren sounded. After the noon whistle scattered all for lunch, three of us crossed the river to eat on benches that faced Route 171. At 12:20, a fellow worker strode down the street from home, entered a bar, and reappeared five minutes later. We clocked in as the whistle blew 12:30. An end-of-day blast sounded at 3:25 so workers could wash up, but most simply waited to punch out at 3:30.

Without meaning to, I absorbed from others how to think about the work. Sweating from a long morning, the three of us sat outside one lunchtime eating peanut butter and jelly sandwiches when a mailman interrupted to chat about the heat and his long rounds. He carried a heavy bag, he said, and his feet hurt. He waited, drew no response, and shuffled off. After he left, my brother and his friend exchanged jokes about the burdens of high pay and good pensions. I nodded: Doyle & Roth work was tougher, the place hotter, and the days more difficult. I lasted two months.

I heard bits and pieces. One foreman dismissed injuries: "put turpentine on it." Another was rightly proud of convincing management to offer, finally, a health-care plan. Before work one day, an early arriver picked off pigeons in the attached storage barn, and the plant closed every year on the first day of buck season. Workers called the place Dirt & Rot.

Water Gas

The local water-energy nexus not only transformed water and coal into electricity but also made water and coal into gas. During the 1902 strike, Scranton Gas and Water Company advertised gas ranges to men as a way to "remove all anxiety as to the *Coal Supply* for your kitchen, and will also save your wife much of the drudgery of housekeeping"; the ad proclaimed that "cooking with Gas is as cheap as coal, is cleaner, and much more convenient." In 1907, the builders of our house added to an attic bedroom a lighting fixture with a choice between manufactured gas and electricity because, I imagine, the choice between them was still an open question. When W. W. Scranton died in 1916, SGW was "the sole manufacturer of illuminating gas here."[12]

Just as they rallied to steam engines in the 1780–1790s, English cotton mill owners successfully experimented with coal gas lighting in 1805. The gas had two advantages over other made gases: saleable byproducts and uniform delivery. Despite its pluses, however, after 1873, coal gas lost out to water gas, a mix of hydrogen and carbon monoxide.[13]

To make water gas, enterprises heated anthracite, introduced steam, and added oil gas to make the flame luminous. Before entering a holding tank, the gas passed from a "wash box" of water to a tar separator. To ensure removal of "all chemical impurities," the system then scrubbed the gas again. By the late nineteenth century, manufacturers took the extracted tar to make aniline dyes.[14]

Concentrated in New York, Pennsylvania, Ohio, and Massachusetts, manufactured gas plants (MGPs) lit and powered many U.S. cities between 1850 and 1950. The widespread availability of affordable anthracite and few by-products helped water gas capture 60 to 70 percent of the U.S. gaslight market by 1912. And MGPs burned a lot of hard coal: in 1927, People's Light Company of Pittston, which served over 4,000 metered locations, annually burned "1700 tons of broken anthracite, about 1,200 tons of rice anthracite and about 375,000 gallons of oil" to make 106,000,000 cubic feet of gas. Low start-up costs, high candlepower, and flexibility of operation added to the appeal of water gas. Beyond material and market advantages, it also offered makers "freedom from labor difficulties, due to [the] smaller number of men required to operate a water-gas plant."[15]

In the 1920s, manufactured gas lost markets to natural gas. Producing a lot of toxins, water gas sparked "frequent . . . citizens' complaints, lawsuits, and legislative action," and scientists worried about its harms to fish, groundwater, and sewage treatment. By the 1930s, upgraded pipelines carried natural gas longer distances to supply more cities, reaching the East Coast after World War II through the Big Inch and Little Big Inch lines. The shift from water gas to natural gas required decommissioning MGPs, but the industry "paid limited attention to the need for thorough remediation," leaving coal tar waste in pits up to 20 feet deep at out-of-service sites. By comparison, natural gas looked clean.[16]

Water Battery

Five energy systems overlap near Forest City: a natural gas pipeline, a wind farm, mined land, and a hydropower project, all surrounded with woods, fields, and farms that feed on sunlight. In 2017, entrepreneurs planned to press the semi-retired Lackawanna back into service for a pumped storage facility designed to plug the river into the Northeast grid. As a hydroelectric site, the Lackawanna would contribute to a renewable energy stream that makes up 7 percent of U.S. electricity production and "56 percent of U.S. renewable electricity." In 2019, the United States produced more power overall from renewable energy than from coal—the first time since 1885—and in 2020, renewables surpassed coal in electricity generation.[17]

In support of renewables, Merchant Hydro sought to place its pumped storage plant close to the Moosic Mountain wind farm, a line of turbines raised in 2003 and visible from downtown Scranton. A water-energy loop, pumped storage generates electricity by capturing the energy of water falling between two reservoirs. A portion of the power then pumps water back to the upper pond. To take advantage of differential pricing, the system moves water at night, when electricity costs less; during the day, when electricity sells for more, it sends energy to the grid. Promoted as clean power, electricity from the project would, as did most coal, leave the region. Water falling near Forest City would generate juice for New York in order to satisfy a state requirement for green energy. Buried in a hard coal past, the current will run north through transmission lines buried beneath the O & W railbed.[18]

Stored water is stored energy. Akin to batteries, pumped storage facilities close supply gaps in the grid, and some can generate reserves on a gigawatt scale. With an upper reservoir larger than several sources of the Lackawanna, the local hydro project will generate 230 MW from water falling over 600 feet between a 260-acre pond and one of 75 acres. By 2050, pumped storage capacity worldwide may jump by a factor of three to five.[19]

Merchant Hydro met skepticism. A Union Dale resident defined the project as "an insult to injury, because . . . the river's upper banks are home to towering culm piles and scars from deep mines and vast surface mining." She wondered, "Are we just a wasteland . . . or are we a real place?" An EPA official warned that the weight of the reservoirs might lead to subsidence of "known and unknown mine voids."[20]

Electric City

Scranton calls itself the Electric City, a nickname proclaimed high above the courthouse square. In 1916, Scranton Electric Company raised the "Electric City" sign atop the Board of Trade Building. Replacing "Watch Scranton Grow," the new lights celebrated the founding of the first electric streetcar system in the U.S. A marketing gimmick, the 60′ × 48′ circular display promoted company efforts to encourage electric signage.[21]

In 1886, a watershed year for local electricity, a spring Authors' Carnival introduced city residents to indoor electric lighting. Eleven booths inside a downtown hall showcased various writers in period dress; each evening featured tableaux from the work of two authors, vividly rendered under colored limelight. In addition to "regular electric lights," illumination also included Edison bulbs, "one in each booth," all powered by a 10-horsepower dynamo. The folks who played parts constituted a who's who among anthracite elites. Mrs. William Connell impersonated Harriet Beecher Stowe; dressed in royal purple, her nephew William Connell portrayed the sultan from *Arabian Nights*.

Later in the year, on December 1, those who heard explorer Henry Stanley recount his travels "Across the Dark Continent" could venture home via the new electric streetcar network.[22]

As happened with water, electricity concentrated into a few hands. Between 1907 and 1917, intense competition forced small generators to merge. In 1907, the national American Gas and Electric consolidated several outfits in Scranton, monopolizing service to the city. By 1910, a company subsidiary, Scranton Electric, had seized control of virtually all power generation in the valley, from Forest City to Pittston. To lure customers away from gas lighting, Scranton Electric pushed the "cleanliness" of electricity, an illusion of clean that persists today, despite most power generation happening at plants that burn coal or natural gas.[23]

Generating electricity with natural gas requires coping with water. In the Marcellus shale region in 2011, American Water, a for-profit public utility, served 12 natural gas drilling companies through 29 points of interconnection. Drawing from former SGW reservoirs, Pennsylvania American Water, an American Water subsidiary, brings water to the gas-fired Invenergy plant in Jessup, which came online in 2019. After a court challenge brought by the Sierra Club, Invenergy abandoned plans in 2017 to daily "discharge hundreds of thousands of gallons [290,000] of industrial wastewater into Grassy Island Creek," which loses its water to subsurface mines.[24]

As does Grassy Island Creek, the 1,500-megawatt Invenergy plant straddles energy eras. Tying the Lackawanna Valley to fracking in Marcellus shale, the facility transforms gas into electricity on former coal lands that suffered a small mine subsidence in 2017. Contributing to the Northeast grid, the plant operates only a few miles from where crews finally extinguished the Powderly Creek mine fire in 2018. Not far in the opposite direction, another still burns in Olyphant; begun in 2004 when someone torched a car, this fire has eaten into the earth over 220 feet and now encompasses seven acres. During the first attempt to kill it, the blaze boiled away 1,600 gallons of water.[25]

Infrastructure: Seen and Unseen

Before 1966, people in the Lackawanna Valley so often saw anthracite infrastructure on the surface—especially breakers and railroads—that it became all but invisible. Unlike their miner neighbors, though, nonminers knew little to nothing about the layers of tunnels beneath them, despite belowground mazes shaping their aboveground world: acid polluted streams, subsidence damaged streets, and mine fires threatened homes. Although coal-mining infrastructure survives on the surface today only in remnants and below mainly in recorded memory, the arrival of natural gas drilling has made above- and belowground newly visible, reminding residents of northeastern Pennsylvania that the extraction and subsidence narratives never ended here. They just opened new chapters.

Oil infrastructure is also highly visible, perhaps to the point of similar invisibility, in gas stations and power lines, highways and parking lots. Just as coal production requires breakers and railroads, oil production requires pumpjacks and truck transport. Natural gas is different: burying pipelines makes its delivery much less apparent. After wells are drilled, gas moves from well bore to pipeline, out of sight, bypassing the surface to spider out to homes, businesses, and processing plants. If natural gas is the second coming of coal, how do I pay attention to it, especially when I can see so little evidence of it?

The afterlife of anthracite laid a pattern for this latest local water-energy nexus, entangling the regional economy again in fossil fuel dependence. Natural gas companies reuse former coal-mining infrastructure to shatter shale and ship methane. In 2012, at the old D & H railyard in Carbondale and near Doyle & Roth, drillers drew water from the Lackawanna River to frack wells in Susquehanna County. With a long history of moving hard coal, the D & H railbed, now a hike-and-bike trail, bears beneath it a pipeline that feeds the Invenergy plant. Headquartered in Pittston, pipeline builder Linde uses as a staging area the empty Vision 2000 Forest City Industrial Park, a leveled culm pile that Hillside Coal Company long ago abandoned.[26]

To ensure that Invenergy receives only methane, water must be separated from it. The continual process begins at the wellhead so that gas moves freely through gathering lines. Before it enters long-distance systems, dehydrating agents or condensation and collection methods remove more water. Compressor stations, which often run on some of the gas they pump, also draw off water and any hydrocarbons that "condense out of the gas stream while in transit." Before gas enters a new pipeline, crews run water through it to check for leaks, a 19-day process for the line to the Invenergy plant.[27]

Pipelines are a safe way to move gas and oil long distances, but any assumption that they offer "perfect containment" defies reality. In 2000, near Carlsbad, New Mexico, a corroded pipeline exploded, killing 12. In 2010, weak regulatory oversight and a power company's "litany of failures" triggered a pipe explosion in San Francisco that killed 8. In the same year, a line in Michigan burst, pouring tar sands oil into the Kalamazoo River; the cleanup neared $1 billion. The United States annually averages "over 125 significant spills of hazardous liquids," dumping "an average of more than 120,000 barrels." Between 2006 and 2015, the number of major incidents jumped 27 percent.[28]

Gas line explosions have happened here, too. On my usual walk to Marywood, I cross Woodlawn Street at Adams Avenue, an ordinary intersection in an old residential neighborhood, unremarkable unless you know its history. In February 1924, a woman reported a water leak at the corner. Just as two employees of Scranton Gas and Water arrived, a pair of explosions killed one, injured the other, and ripped open a hole 75 feet deep and 70 feet wide. Officially the cause went undiscovered, but even Pennsylvania Coal Company, whose men mined beneath the property, blamed the blast on a mine subsidence that ruptured a gas line.[29]

The story of this intersection echoes more recent ruptures in time and space. As multinationals pockmarked Pennsylvania with boreholes in 2007, others crisscrossed the commonwealth with pipelines to deliver gas to homes and businesses worldwide. Workers deepened this web in 2008–2016 by building "more than 450 natural gas compressor stations and processing plants" and in 2011 by laying a larger pipeline next to the existing one that extends east from the Stillwater reservoir. This enlarging infrastructure, rural and urban and much of it underground, led me to adjust my sight lines. As I came to know the state's extraction story better, I began to write in a more fluid form, which helped me see beyond surface grids to discover the water that buoys fossil fuel production. Without an adaptable narrative, I risked turning local spots *of* time into spots *in* time, which would have only suggested desiccated questions—and answers—about this coal region.[30]

10

Fracking Water

If you wanted answers in ancient Greece, you went to Delphi. Apollo, god of sun and poetry, spoke the truth there through a succession of priestesses named Pythia. To ready herself to channel the immortal, she washed in spring water, perched on a tripod above a chasm, and inhaled sweet vapors. Inspired to a frenzy, she raved as the god, whose language priests translated for petitioners, who mainly learned that gods reveal ambiguous truths.[1]

Pythia's source of inspiration may have been natural gas. The shrine at Delphi sat atop an active fault, whose fractures allowed gas and water to find the surface. The spot exhaled carbon dioxide, methane, and benzene, an aromatic carbon that can impair vision, shorten breath, and destabilize emotions. Unlike Pythia, multinationals these days declare gas the answer rather than the inspiration for uncertainty.[2]

Promoters tout natural gas as a bridge between a fossil fuel past and a green energy future. Arguing to buy time for a global energy transition, they claim that until wind and sun can power modern systems, gas is the best bet to cut carbon emissions, short of shutting down industrial economies. But the world is running out of time; scientists believe that we are now experiencing a "planetary emergency" that "requires an emergency response." A half answer to questions about energy and climate, natural gas—a fossil fuel—may be a bridge to nowhere.[3]

Undermining arguments that natural gas can bring the world across the energy gap, in 2011, Robert Howarth and Anthony Ingraffea discovered that fracking releases twice as much methane into the atmosphere as conventional drilling. Concluding that gas was little better than coal, they claimed that to define it as a bridge fuel is to put faith in an industry myth. But by 2014, when Ingraffea called out methane leaks from failed cement casings, the industry had fracked at least 100,000 wells nationwide, entrenching the myth as fact.[4]

Methane is a serious problem. A greenhouse gas, it's much more potent in the short run than carbon dioxide. Over a 100-year stretch, it traps 34 times more heat than CO_2; over 20 years, it traps 86 times more. With effects of climate change cascading upon us, 20 years is no time. And because fugitive emissions are tough to detect, methane emissions may be higher than previously thought.[5]

Mining Water

Punching holes through time, crews drill a conventional well straight into the earth. In unconventional drilling, they do the same but then turn the bit at a right angle to bore through a single era. Although few companies drilled horizontally in 2002, by 2012, more than half did. Driving the increase: the discovery in 1998 that production soared at conventional wells when drillers pumped down water to fracture shale, a process called hydraulic fracturing, or fracking. Soon used in horizontal drilling, fracking helped boost U.S. gas production from 320 billion cubic feet in 2000 to 15.8 trillion in 2016.[6]

Hydraulic fracturing has become commonplace. Between 25,000 and 30,000 wells get fracked each year in the United States, with just over 80 percent producing oil and the rest natural gas. In 2011, an expert estimated that a mature Pennsylvania gas field would count between 60,000 and 170,000 wells; by 2017, 11,000 unconventional ones had been drilled. A result: in 2018, earlier than previously estimated and mainly due to fracking, the U.S. surpassed Russia and Saudi Arabia as the world's top oil producer, a first since 1973, but a dubious distinction in an era of climate change.[7]

To free gas, drillers hammer shales with high-pressure injections of water and sand that they lace with toxic chemicals to accelerate the flow. After the rock breaks, sand holds open the fissures, allowing gas to find the well bore, which it follows to the surface. A typical frack breaks up about "one hundred million square feet—or the floor-space equivalent of about thirty-five giant malls." And one well may get fracked as often as 15 times, making "well bore integrity . . . paramount for preventing groundwater contamination."[8]

Fracking shales is water intensive. The amount varies by location, but overall, the practice uses about 50 times more water than conventional wells. Although the national average to frack a gas well is 1.7 million gallons, in the Marcellus shale of Pennsylvania, it takes between 3.8 and 5.5 million gallons. Some "monster fracks" in Texas suck in over 40 million gallons. In contrast to shale, fracturing a well in a coal bed formation takes only 50,000 to 350,000 gallons. Unfortunately, about half of shale oil and gas lies under water-stressed regions.[9]

Frackers not only pump a lot of water into the earth, but they also collect a lot of toxic flowback, making wastewater the industry's biggest output. In the Marcellus field, about 10 percent of water sent underground returns to the surface, with the average well spewing 1.25 million gallons, which is a third flowback, half brine, and the rest drilling fluids. In 2008–2011, wells in Pennsylvania coughed up 1.3 billion gallons of wastewater, "enough to cover Manhattan three inches deep." In 2015 alone, fracked wells poured back about 1.7 billion gallons. Although fracking generates 10 times the wastewater of conventional drilling, it frees 30 times more gas.[10]

Wastewater must go somewhere. At first, some spilled—or was dumped—into streams and rivers, but much of it went to sewage treatment plants. In 2010, Pennsylvania updated its Clean Streams Law to require that wastewater be treated to meet federal and state drinking water standards, a much stricter measure than for other industries. Despite this, in 2017, Penn State researchers discovered that treated wastewater left high concentrations of toxins in sediments as far as 12 miles from treatment plants. Uproar over wastewater led the industry to experiment with closed-loop recycling, which kept wastewater on site, but most companies turned to injection wells, often in other watersheds, where it went back underground.[11]

Beyond creating pools of poison, putting flowback underground has increased earthquake activity near injection sites. After 2009, Oklahoma, a state with thousands of injection wells, became 100 times more seismically active than it had been, suffering its largest tremblor, a 5.8 quake, in 2016. Although few quakes in Oklahoma have been linked to fracking itself, some in Ohio have been attributed to it.[12]

Just as anthracite owners mined culm banks, some investors see a gold mine in gas-industry wastewater. The Permian Basin daily deals with 1,000 Olympic-style swimming pools of it, which will only increase because the field's oil production might jump to five million barrels a day by 2023, more than Iran's everyday output in 2018. Some Permian wells "produce 10 times as much water as they do hydrocarbons," and handling it accounts for 25 percent of expenses per well. To capitalize on the problem, wastewater manager WaterBridge uses a network of pipelines to collect and recycle 600,000 gallons of the waste every day. Outfits like WaterBridge clean wastewater to sell it back to companies that had paid them to get rid of it.[13]

Mining St. Peter

A place is many places. I often see scores of covered hoppers parked at the former Lackawanna Railroad Station, near the University of Scranton and not far from the Lackawanna Historical Society. Passing behind the Scranton iron furnaces, engines of the Delaware-Lackawanna Railroad push these cars of silica sand north past Steamtown National Historic Site. Trundling through Carbondale, the hoppers parallel the Lackawanna River and pass Pleasant Mount Welding before they reach the old D & H railyards. Next to Carbondale Ready-Mix, a concrete maker, and only steps from Doyle & Roth, teams top off tractor trailers that haul the sand to drill pads in Susquehanna County. The site helps make sand mining a $70 billion global business.[14]

Extending the effects of Marcellus drilling, the sand originates in the Driftless Area, parts of Wisconsin and Illinois untouched by the Laurentide ice sheet. Although never buried in glacial till, the region funneled meltwater into a Paleozoic sea, depositing the sand, called St. Peter sandstone. Used to prop open rock fissures, the sand can be customized by gas field and well. St. Peter sand particles are 99.8 percent silica, mainly round, and crush resistant. Selling in 2013 for about $55 per ton, much Midwest sand comes from surface mines.[15]

After quarrying, the sand must be processed, which can require as much as two million gallons of water each day. A washing in 4,000–6,500 gallons per minute removes impurities, with wastewater going to settling ponds. After drying and sizing, the sand gets shipped to gas fields. To frack the average well in 2014, crews needed between 40 and 50 railcars of sand, about 4,000 to 5,000 tons; some wells sucked in 9,000 tons, or nearly 100 cars. The Delaware-Lackawanna Railroad, which usually moves sand to Carbondale in 100-car trains, broke a company record in 2018 for freight hauling due to increased demand for frac sand.[16]

Heat and Light: Water and Ground

Jennifer Haigh's *Heat and Light* (2016) recounts the advent of natural gas drilling in a former coal town in western Pennsylvania. Probing subtexts of today's fossil fuel transition and portraying people who are too often scrubbed from industry narratives, the novel stresses uncertainty, which may account for its fractured form. Refusing resolution, *Heat and Light* uses irony, paradox, and juxtaposition to expose cracks in a community divided over fracking.[17]

Heat and Light explores real-world debates about whether fracking affects groundwater. During their first dinner date, Herc, a Stream Solutions foreman, assures Jess, pastor of Living Waters Church, that drilling does not endanger local water. In reply, she points to faucets on fire out West, a fact that he labels propaganda from "environmental nut jobs trying to scare you" (129). In discussing the "problem of flowback" (223) sent to sewage treatment plants, geology professor Lorne Trexler explains that frackwater mixed with bacterial disinfectants creates carcinogens that end up in waterways, a circumstance, he notes, made legal through the Halliburton loophole, which "exempts fracking fluid from the Clean Water Act" (224). Trexler's talk convinces Shelby Devlin that methane in her well water has poisoned her daughter, Olivia (226).[18]

Shelby's fear shows that arguments about water contamination are not academic abstractions. Uncertainty about it shapes the Devlins' everyday lives. At first dismissing Shelby's complaint that their water smells (213), her husband, Rich, knows she's right after he sniffs something like lighter fluid in the shower (258). When he confronts Herc (259), the driller not only tells him to test the water and call the company, Dark Elephant, but also that doing so amounts to "wasting your time" (260). A letter from the state department of environmental protection acknowledges methane in the Devlins' water but does not link its presence to drilling or mention chemicals (325–326). After a second test confirms methane migration, Dark Elephant claims that "the water was dirty to begin with" (360). Frustrated, and with few options, Rich buys water at Walmart for drinking and cooking (323), but when he arrives home one day, he discovers that a worried Shelby has been using it to wash dishes and do laundry, drawing down the family's fragile finances (326–327). Admitting no fault but offering "*a good neighbor gesture*," Dark Elephant later supplies the Devlins with a water buffalo, a temporary fix to a lasting problem (424).

The uncertainty around fracking also unsettles the Devlins' land, physically and legally (216). After cutting an access road, a crew transforms their pasture into a "razed and flattened" well pad (193), an updated version of coal-era "strippins—high and sloped, machine-graded," evidence that time has not passed but cycled round (358). Knowing that he cannot farm without clean water, Rich assumes that he can sell the land, but his attorney points out that he may not have clear title to it because he signed a lease that allows the company to place a lien on the property. Without clear title, he can't sell (362). The land remains in limbo: it may or may not be his, and it may or may not be a farm.

Cement Seals

Concrete is the only substance humanity uses more than water. Third behind China and the U.S., the concrete industry annually pours nearly three billion tons of carbon dioxide into the atmosphere. Emissions run high because making its binder burns fossil fuels; to produce a ton of cement is to burn about 0.1 ton of coal, making cement making the third major market for coal, behind electricity and coke production.[19]

Cement is not a foolproof seal, but oil and gas companies want us to believe that fracking will not contaminate groundwater because drillers encase wells with it. Besides, they say, frac water goes so far into the earth that there is no chance that toxins can make their way toward the surface. But when drillers cut through groundwater supplies, breaks in cement often mean tainted water. Triple casings may work in the short term, but in the long run, casings will almost certainly crack: cement fails, steel rusts, the earth moves. The more wells that get fracked, the more chances that fresh water gets poisoned.[20]

The state's extraction history cautions against faith in sealing wells. A "wild card" for drillers, former mines stretch between shale and surface, making them funnels for fugitive gas to reach groundwater. Not only are parts of Pennsylvania a network of abandoned mines, but some places are also shot through with old oil and gas wells, as many as 1,600 unplugged. The state may have around a half million more abandoned wells, but no one knows for sure because nobody kept records before the mid-1950s.[21]

Industry practices work against sealing. To create a proper one, a crew needs to trust the material and take time to do it right. Unlike in building bridges and roads, though, in drilling for oil and gas, "the cement job is not regulated, not even the cement composition." Pressed to maintain production, companies sink wells as fast as possible because each one's output decreases right after it's drilled. Coming at the end of drilling, cementing most often happens when "everyone's wanting to move on to the next," an attitude that encourages haste. As an oil and gas worker observes about companies, "Here's how they operate: 'At this point, we don't care. If there's a problem, we'll fix it later.'" Add to this that in 2008, in the early headlong rush to capture shale gas in Pennsylvania, engineers admitted that they were unsure how shale production worked, and most acknowledged that they knew little about "reservoir drainage."[22]

Pennsylvania has accused companies of not sealing wells properly. Between 2009 and 2014, about 12 percent of fracked wells in Pennsylvania had structural problems. In 2009, the Pennsylvania Department of Environmental Protection (DEP) reported that Cabot Oil & Gas had not properly cemented at least six wells. In the first nine months of 2011, the DEP handed out "eighty-nine citations for faulty casing and cementing practices" across the state. In 2012, the state required Royal Dutch Shell to fix cement work on a new well in Tioga County after a "geyser of methane-laced water . . . shot twenty to thirty feet into the air" from a nearby undocumented one. In 2020, the DEP ordered Range Resources to fix a cement casing in Lycoming County that the state claimed had been leaking methane into streams and groundwater since 2011.[23]

Leaks discovered are not always leaks rightly reported. In 2017, DTE Energy failed to mention to emergency management officials that gas had leaked from a compressor station not far from Lanesboro, Susquehanna County. Despite state rules that require companies to report leaks to first responders, DTE notified only the DEP and the U.S. Pipeline and Hazardous Materials Safety Administration. Word of the leak reached the public only by accident after the Associated Press reviewed calls to a U.S. Coast Guard hotline. In this case, "leak" may be an understatement. In 2014, the average compressor station released 107 tons of methane; in 2015, this dropped to 97.5 tons. The DTE station pumped out 200 tons in two hours.[24]

Pennsylvania worries about seals because bad cement jobs have led to "heart-breaking stories of homeowners, and burning faucets." In 2007, a casing failure in Ohio led to a house explosion. In 2011, a leaking cement seal in Jefferson County, Pennsylvania, threatened the underground aquifer that supplied a nearby town. In the same year, cracked cement allowed gas to contaminate groundwater in Bradford County.[25]

At some point, the number and extent of sacrifice zones may dwarf "safe zones." Stepped-up drilling for oil and natural gas has already led to "major industrialization of both rural and urban parts of the United States." Drilling 5,000 wells between 2005 and 2011, companies sank 80 percent of them in rural counties, transforming them into industrial sites. By 2017, over 17 million Americans lived within a mile of an active well, with over 4.5 million affected in Texas alone, including residents of Fort Worth. As the "most densely populated urbanized mining region" in the nation, the northern anthracite field has experienced what this means.[26]

Scientists say that time to act about climate change is running out, but moving fast is itself a problem because some mistakes can't be called back, especially when good water may go bad. Going all in on natural gas is not the answer. Seeing time as money may accelerate proposed solutions, but getting in step with the natural world means slowing down to work in rhythm with air, light, and water. Acknowledging limits does not restrict imagination; think about the inexhaustible nature of the haiku, the sonnet, and the villanelle—and soil, field, and forest.

Heat and Light: Seeing and Responsibility

Fracking raises questions about responsibility. Cement may suggest certainty—the seal will keep toxins at bay—but a small crack can trigger widespread groundwater contamination, disrupting the gas-as-clean-energy narrative. As Herc, the drilling foreman, drives by the Devlin home, he wonders aloud whether a bad cement job might have leaked methane to their well. The cement crew he hired was new to him, and he didn't supervise the job, so he can't vouch for it (342). When coworker Mickey tells him that it's none of his business, Herc replies, "I'm going to feel responsible, even if it isn't my fault" (342). But if he had watched the crew pumping cement into the hole, he muses, he would have seen nothing, and, he admits, "What actually happened downhole was anybody's guess" (342). Taking orders from Dark Elephant, which had chosen "a cheaper and faster solution" (342), Herc had little control of the process anyway. Ultimately, he takes Mickey's advice: he says nothing.

Signaling that fractured knowledge complicates ethical choices, *Heat and Light*, as author Jennifer Haigh says, reveals "how our lives are shaped by large forces—economic and political—we understand incompletely or not at all." To enact this idea, she places readers in the position of making connections that her characters do not. Unlike the Devlins, who focus on how the oil and gas industry affects their everyday lives, readers come to know Carbon Township in relationship to events widely separated in time and space: the 1973 oil shock (135), the 1979 Iranian revolution and hostage crisis (168), Chernobyl (307), the First Gulf War (217), 9/11 (323–324), and wars in Iraq and Afghanistan (187). The Devlins' perspective warns us about what we all experience to some degree, "You kind of stop looking. . . . It's the fundamental problem of a life lived in one place: sooner or later, everything becomes invisible" (235). If seeing is a form of responsibility—and I believe it is—the novel prompts readers to look at and respond to their own, and their community's, scarcely understood addiction to fossil fuels, if only to decide not to act.[27]

Dimock

Susquehanna County, New Year's Day, 2009: a water well explodes in Dimock, bringing my full attention to drilling. Under the well's concrete cover, methane had built up and ignited, blowing off the cap. Tests at the site found high levels of manganese, lithium, arsenic, propane, and uranium. Like others who saw symbolism in the event, I took it as a sign to watch water. As the place quickly became a center of international debate, I saw that people in Dimock were coming to know water in new ways.[28]

Methane from a Cabot Oil & Gas well popped the cap, but the company denied responsibility. Betting its future on the Marcellus play, Houston-based Cabot poured $600 million into its shale operations, mainly in Susquehanna County. Between 2006 and 2015, the company drilled nearly 500 wells, but not all were drilled well. After methane seeped into Dimock groundwater, the DEP forced the company to provide water buffalos, but in 2011, the state allowed it to halt deliveries, leaving some residents to survive since then on bottled water. In 2012, a federal public health report found that 27 private water wells in Dimock contained "threatening levels of chemicals" and 17 contained "explosive levels of methane." After declaring the water safe in March 2012, the EPA announced in July that "residents have now or will have their own treatment systems that can reduce concentrations of those hazardous substances to acceptable levels at the tap." Little comfort, the phrase "acceptable levels" led me to wonder, acceptable to whom?[29]

Water problems continued, making Dimock a focus of controversy. In 2015, Cabot racked up the second-highest number of state rule violations, behind only Chesapeake Energy. In 2020, the state attorney general's office charged Cabot with 15 environmental crimes, including improperly secured well casings and methane seepage into sources of drinking water. According to the grand jury report, a DEP investigation discovered that migrating methane originated deep in the earth, and not in shallow formations, as Cabot argued. Methane from deep down means that drilling brought it to the surface, making bad casings and bad drinking water a much more likely cause and effect. In a gesture to the commonwealth, the epigraph of the grand jury report repeats Article 1, Section 27 of the Pennsylvania Constitution, which guarantees state citizens environmental rights, among them access to clean air and water, a provision arising in response, in part, to the Knox mine disaster, the Centralia mine fire, and the Glen Alden fish kill.[30]

Despite grand juries and constitutional guarantees, the gas industry worries little about oversight. Not very aggressive in tracking methane migration problems, Pennsylvania "does not report how many complaints about gas in water have been called in, or how many homes can no longer use their water wells." Employing too few inspectors, the state relies on industry records, which are often "filed away or incomplete" and written long after the incident. Oversight has been so scarce that a Penn State Cooperative Education representative asked residents "to keep an eye" on drilling. Adding to company powers, the state legislature refuses to impose a severance tax on the industry.[31]

Tensions between money and water in Dimock also flared elsewhere. In its decision-making about fracking, the Delaware River Basin Commission (DRBC) had to choose between property rights and safe drinking water for

New York, Philadelphia, and people in between. After the DRBC placed a moratorium on drilling in the watershed in 2010, landowners in Wayne County sued. Seven years later, the DRBC voted 3-1-1 to draft resolutions to ban fracking permanently in the basin, which includes most of Wayne County. A few months before this decision, the Pennsylvania legislature signaled its stance when it cut its contribution to the DRBC in half, to $217,000, less than a quarter of what the state should provide. The DRBC made the ban official in 2021. Still unsettled, the landowners' suit hinges on whether fracking is a "project."[32]

Behind debates about natural gas stands anthracite history. Arguments about property rights and clean water echo Scranton-based court decisions about local mining. In *Pennsylvania Coal v. Sanderson* (1886), the state supreme court affirmed a "'property right' to pollute" by making it legal for acid mine drainage to poison a private pond. In *Pennsylvania Coal v. Mahon* (1922), a case about mine subsidence, the U.S. Supreme Court extended the definition of government seizure of property to regulations. In 2020, PennFuture and Conservation Voters of Pennsylvania ran a full-page ad in the *Scranton Sunday Times* to encourage readers to urge elected officials to vote against state subsidies for the petrochemical industry. Remembering that "the first anthracite mine was established near Pittston in 1775," the ad drew a parallel between hard coal legacies of "barren black hills, orange rivers, abandoned factories, black lung and cancer" and the natural gas industry's "toxins like benzene, formaldehyde, toluene and a host of other pollutants." The ad wondered why politicians would "want to give away billions in taxpayer handouts to some of the richest corporations in the world, instead of investing in cleaner industries." A reasonable question. Any reasonable answer should, at minimum, protect the commonwealth's water.[33]

11

Restoring Water

As a junior at the Jesuit-run University of Scranton, I took a philosophy course on the pre-Socratics that met in St. Thomas Hall. Thirty minutes into each class, church bells chimed the noon hour, ringing out the Angelus, which meant that we couldn't hear the teacher, who paused. Listening, he waited for the last echo of the last note to die before resuming. Here I first heard about Thales.

Aristotle named Thales the original philosopher because he attributed natural phenomena not to gods but to a material cause, water: "*all whatness is wetness.*" The first scientist, Thales wondered about change: when water freezes, is it still water? When it evaporates, does it just disappear? He also claimed that a lodestone—a magnet—moved because it had a "soul," challenging the belief that an outside force—a god—touched it. His home, Miletus, in present-day Turkey, later had to be abandoned because its river, the Meander, silted the city's harbor.[1]

Thales served as the model for the central figure in Aesop's fable "The Astrologer Who Fell into a Well," which Plato first recorded in *Theaetetus*, an exploration of the nature of knowledge. Perhaps about the first absent-minded professor, the story depicts an astronomer gazing at the stars so intently that he stumbles into a hole; bruised, unable to escape, he shouts for help. The servant girl who brings him to the surface tells him that if he must study the heavens, he should keep an eye on the ground. Watchful of a warming atmosphere, the industrial world is like Thales, about to trip into another well, a gas well. To skirt the trap, we should keep both eyes on land and water.[2]

Men mined rivers and streams. In the early twentieth century, an anthracite navy prowled the Susquehanna, searching for what may have been nearly one billion tons of anthracite in 1904. For 100 miles, Scranton investors sailed the river to retrieve the coal they had let wash away at home.[3]

Density determined where coal would settle. Using ash poles 18–20 feet long, diggers poked at the river bottom; if they heard an ice crystal sound, they knew they had anthracite. Operating April to October, the coal fleet brought up 12.6 tons per man per day in 1940. Rivaling early washery output, dredgers took two million tons in 1919 and 1920. In these years, river coal accounted for less than 1 percent of anthracite mined; by 1940, output had doubled, to nearly 2 percent. In the mid-1950s, river profits dried up after the state prohibited collieries from pouring wastewater into streams.[4]

Individuals and companies launched boats, making the Susquehanna one of the few rivers in the world to float a successful coal business. In 1914, 47 outfits vacuumed up 150,000 tons, including Harrisburg Light Company, which collected 29,000 tons. Dredgers also supplied Wilkes-Barre Railway, Sunbury Water, and Susquehanna Woolen Company. In Pittston in 1915, Citizens' Electric, which relied on a culm bank at the Clear Spring colliery in winter, fed boilers in summer with coal pumped from the Susquehanna. In the 1920s, river coal heated Harrisburg School District, and in 1996, Bucknell University may have bought the final commercial load. One of the last river diggers, who retired in 1996, sold coal to Pennsylvania Power & Light.[5]

Some dams doubled as mines. In 1914, two dredging operations recovered 48,000 tons of steam coal from behind levees on the Lehigh River. In the same year, the Laurel Line in Scranton powered trolleys with coal and water mined from behind a dam on Roaring Brook; removing the coal also purified the pond's water, which the company piped to the boiler's condenser. In 1925–1953, Holtwood Hydroelectrical Plant burned what accumulated behind Holtwood Dam, as much as 10 million tons.[6]

Renewal requires rethinking, not returning. The gods suggest so in offering new beginnings after water wipes the world clean. I discover this in the *Epic of Gilgamesh*, in Greek mythology, and in the Hebrew Bible, which all recount flood stories. The Greeks say that Deucalion, son of Prometheus, and his wife, Pyrrha, survive a world flood due to their virtue. As the water recedes, they find themselves on a mountaintop, where a goddess tells them to throw the bones of their mother over their shoulders. Astonished, the couple assumes that this would insult the dead, but they then realize that they are to toss stones behind them. As they do, the pieces turn into people, repopulating the earth and signaling a new start.[7]

I live in the aftermath of mining. But aftermath also means second growth, reminding me that the valley was not killed so much as altered. In "aftermath" I hear renewal because the word denotes not only the period right after a ruinous event but also its effects and what comes next, new life. An industrial site can never be made whole if doing so means returning land to an original state, which is always a legacy of another era: the Appalachians rose and eroded four times. Wholeness is health, so a place to start the healing is to stop the abuse, which signals accepting responsibility for the hurt. People should care, if only because we can. If not us, who?[8]

Ending mining began a revival of the Lackawanna River. Restoring it to health, however, was to see it anew, up close, not as it was, whatever that might have been. In the early 1960s, those who cared decided to get dirty pulling trash from the channel. In 1987, the grassroots group formalized their efforts to create a right relationship with water and began reintroducing the river to the community that had ill-used it. Recovery took time, and people had to pay attention to help the process along, but the labor paid off. Once dead water, the river now boasts trophy trout fishing and in 2020 was voted the best river in Pennsylvania. Work continues, though: along with typical pollutants—road salt, yard waste, tossed garbage—the water still bears "heavy metals, sulfides, and sediments from thousands of acres of mine spoil piles."[9]

In 2016, the river's keepers renamed themselves the Lackawanna River Conservation Association (LRCA), a move meant to acknowledge that the wholeness of the river is about more than the waterway. It's about the watershed, 350 square miles of scarred ground that cycles water within a world system. To know better "how we impact the flow of water through our lives," this wider embrace confronts head-on "long-term issues of sustainability and resiliency in the face of ongoing climate change." Inviting others to act, the LRCA assembled a band of water watchers, the Lackawanna River Watershed Conservation Corps. As they keep an eye on land and water, natural gas drilling threatens the headwaters, and the river's last three miles run lifeless. I live in between, a place sodden with uncertainty.[10]

Although coal mining is history here, its memory is close enough to be felt but distant enough to require research to recall. Fewer and fewer remember the mixed emotions people knew when deep mining ended. In Forest City in 1937, crews dismantled a D & H breaker as the last loaded car surfaced at the Stillwater mine. Seeing this, former miner Anton Zaitz commented, "It was as if an invisible hand suddenly drew the curtain in the middle of a live performance. . . . What happens now?" Answers swept through town: no jobs, no money, no prospects . . .[11]

Witnessing economic collapse, Zaitz saw the land anew: "I experienced the feeling of walking through an abandoned, fire-ravaged country. Everything was dug up and ugly looking. No attempt was made to return the land to its former beauty. Huge mounds of raw dirt and rock . . . looking like a bombed-out scene on some distant battle field." Despite knowing that the moment meant further impoverishment for many, he expressed little regret at the loss of the industry: "With a thumbs-down gesture and a wave of my arm, I exclaimed: 'To hell with you worked-out mines; you won't be missed. Let the cold waters flood your abandoned chambers! Enough lives have been sacrificed on your altar of profit for the absentee owners.'" Now little more than a third of its 1930s population, Forest City, nearly treeless, lives with legacies of mining, which are mainly absences: stripped land, no industry, few young people, and even fewer good-paying jobs.[12]

Locals have rallied to protect remnants of the region's water history. In the 1830s, the D & H dammed a wetland to create a 90-acre reservoir for the canal, later named Hankins Pond, a place just west of our old farm. In 1918, the state bought the site, along with other lakes in Clinton and Mount Pleasant Townships: Belmont, Long Pond, Miller, Swamp, and White Oak. For decades, the state hatchery in Pleasant Mount raised fish at Hankins, but in 2013, they abandoned the practice and drained the pond. In 2018, the Pennsylvania Fish and Boat Commission moved to tear down the dry-stone dam, which the agency defined as a high-hazard structure that it claimed had deteriorated enough to threaten flooding downstream. Wayne County obtained an injunction to stop the dismantling, and in 2019, the commission sold the site to the county, which purchased the property in part using state fees collected from the natural gas industry.[13]

Prior to industrialization, CO_2 levels in the atmosphere hovered between 250 and 290 parts per million; between 1850 and 1958, CO_2 jumped from 280 to 320 ppm, and in 2015, it shot to over 400 ppm. Today: 421. The most widely accepted readings of atmospheric CO_2 come from a monitoring station in Hawaii. In 1958, Charles David Keeling and Roger Revelle, scientists with Scripps Institution of Oceanography, established the facility atop Mauna Loa to be sure its equipment would not register emissions from local industries. Since then, the station has published data in the form of the Keeling Curve, a sawtooth graph of CO_2 concentrations in the Northern Hemisphere. The graph's shape moves with the seasons.[14]

Charles Keeling grew up in Scranton. His childhood home is within walking distance of the old homes of journalist Jane Jacobs and President Joe Biden, a coincidence that reminds me that urbanization, global warming, and international politics will soon force us all to get right with water. To see it anew, however, means more than stats and graphs. It means reminding ourselves just how wet the world is.

Hearing water in our stories might help us revise how we think about it. Making old ideas new and familiar ones strange, storytelling reimagines metaphors, recharging our thinking and revising what we make manifest in the world. The stories that water tells can help us think a better world into being. Re-storying is restoring.

Acknowledgments

Mary Ann Savakinus, director of the Lackawanna Historical Society (LHS), Sarah Piccini, LHS associate director, and Olivia Bernardi, LHS creative director, responded patiently to my many requests for sources and advice. I thank them not only for their outsized help with this book but also for their extraordinary work on a daily basis to keep local history alive and visible.

Several people read drafts with obvious care and attention. Bernie McGurl, who responded to a full draft, made many helpful suggestions. Alex Vari read parts of the manuscript and offered much good advice. The anonymous readers who responded to the manuscript made excellent recommendations. You have all had major effects on the final draft. Thank you.

Four organizations granted me permission to reproduce photos from their collections. I'm grateful to John Fielding, Anthracite Heritage Museum; Mary Ann Savakinus, Lackawanna Historical Society; Bernie McGurl, Lackawanna River Conservation Association; and Joe Butkiewicz, *Scranton Times-Tribune.* I thank Erin Greb for making such great maps. I appreciate the prompt response of John Inners to my request for a source.

The staff of Temple University Press guided me through the publication process with unmatched professionalism. I thank Aaron Javsicas, editor-in-chief, for his interest in the project, and I am grateful for the tireless work of editor Shaun Vigil and coordinator-of-all-things Will Forrest. I am indebted to Jamie Armstrong, who flawlessly managed the book's production, and copy editor Laura Larsen, who paid meticulous attention to every aspect of the manuscript.

My wife, Bridget, read drafts and listened patiently as I thought out loud about water and local history. Thank you, Bridget.

Glossary

The definitions here are quoted from Thomas J. Foster, *Coal Miners' Pocketbook*, 11th ed.

Anthracite: Coal containing a small percentage of volatile matter.
Barrier pillar: A solid block or rib of coal, etc., left unworked between two collieries or mines for security against accidents arising from influx of water.
Borehole: A hole made with a drill, auger, or other tools, in coal, rock, or other material.
Breaker: In anthracite mining, the structure in which the coal is broken, sized, and cleaned for market. Known also as *Coal Breaker*.
Breaker boy: A boy who works in a coal breaker.
Bucket pump: A lifting pump, consisting of buckets fastened to an endless belt or chain.
Carbon: A combustible elementary substance forming the largest component part of coal.
Cave, or **cave in**: A caving-in of the roof strata of a mine, sometimes extending to the surface.
Coffer dam: An enclosure built in the water, and then pumped dry so as to permit masonry or other work to be carried on inside of it.
Colliery: The whole coal mine plant, including the mine and all adjuncts.
Creep: The gradual upheaval of the floor or sagging of the roof of mine workings due to the weighting action of the roof and a tender floor.
Culm: Anthracite-coal dirt.
Culm bank, or **culm dump**: Heaps of culm now generally kept separate from the rock and slate dumps.
Dead work: Exploratory or prospecting work that is not directly productive; brushing roof, lifting bottom, cleaning up falls, blowing rock, etc.
Face: (1) The place at which the material is actually being worked, either in a breast or heading or in longwall. (2) The end of a drift or tunnel.
Fault: A fracture or disturbance of the strata breaking the continuity of the formation.

Flush: (1) To clean out a line of pipes, gutters, etc., by letting in a sudden rush of water. (2) The splitting of the edges of stone under pressure. (3) Forming an even continuous line or surface. (4) To fill a mine with fine material. Sometimes called slush.

Gangway: The main haulage road or level.

Goaf, or **goave**: That part of a mine from which the coal has been worked away, and the space more or less filled up with waste.

Jig: (1) A self-acting incline. (2) A machine for separating ores or minerals from worthless rock by means of their difference in specific gravity; also called *Jigger* or *Washer*.

Jigging: Separating heavy from light particles by agitation in water.

Lode (Cornish): Strictly a fissure in the country rock filled with mineral; usually applied to metalliferous lodes. In general miners' usage, a lode, vein, or ledge is a tabular deposit of valuable minerals between definite boundaries.

Measures: Strata.

Mining: In its broad sense, it embraces all that is concerned with the extraction of minerals and their complete utilization.

Mining engineer: A man having knowledge and experience in the many departments of mining.

Openings, an opening: Any excavation on a coal or ore bed, or to reach the same; a mine.

Pillar: (1) A solid block of coal, etc. varying in area from a few square yards to several acres. (2) Sometimes applied to a timber support.

Place: The portion of coal face allotted to a hewer is spoken of as his "working place," or simply "place."

Pothole: A circular hole in the rock caused by the action of stones whirled around by the water when the strata was covered by water. They are generally filled with sand and drift.

Prop: A wooden or metal temporary support for the roof.

Quarry: (1) An open surface excavation for working valuable rocks or minerals. (2) An underground excavation for obtaining stone for stowage or pack walls.

Reservoir: An artificially built, dammed, or excavated place for holding a reserve of water.

Rob: To cut away or reduce the size of pillars of coal.

Robbing: The taking of mineral from pillars.

Seam: Synonymous with *Bed*, *Vein*, etc.

Screen: (1) A mechanical apparatus for sizing materials. (2) A cloth brattice or curtain hung across a road in a mine, to direct the ventilation.

Shaft: A vertical or highly inclined pit or hole made through strata, through which the product of the mine is hoisted, and through which the ventilation is passed either into or out of the mine. A shaft sunk from one seam to another is called a "blind shaft."

Shaking screen, or **shaker**: A flat screen, often inclined, which is given an oscillatory motion and is used for sizing coal.

Shale: (1) Strictly speaking, all argillaceous strata that split up or peel off in thin laminae. (2) A laminated and stratified sedimentary deposit of clay, often impregnated with bituminous matter.

Sizing: To sort minerals into sizes.

Slate: (1) A hardened clay having a peculiar cleavage. (2) About coal mines, slate is any shale accompanying the coal, also sometimes applied to bony coal.

Squeeze: See *Creep*.

Stratum (plural, **strata**): A layer or bed of rocks, or other deposit.

Tunnel: A horizontal passage driven across the measures and open to day at both ends; applied also to such passages open to day at only one end, or not open to day at either end.

Turbine: (1) A rapidly revolving waterwheel, impelled by the pressure of water upon blades. (2) A similar type of power generator propelled by steam or air.

Unconformability: When one layer of rock, resting on another layer, does not correspond in its angle of bedding.

Vein: See *Lode*. Often applied incorrectly to a seam or bed of coal or other mineral.

Washing apparatus, or **washery**: (1) Machinery and appliances erected on the surface at a colliery, often in connection with coke ovens, for extracting, by washing with water, the impurities mixed with the coal dust or small slack. (2) Machinery for removing impurities from small sizes of anthracite coal.

Waste: (1) See *Goaf*. (2) Very small coal or slack. (3) The portion of a mine occupied by the return airways. (4) Also used to denote the spaces between the pack walls in the gob of longwall working. (5) Refuse material.

Watershed: The elevated land or ridge that divides drainage areas.

Waterwheel (English): Overshot, undershot, breast wheels, etc. A wheel provided with buckets, which is set in motion by the weight or impact of a stream of water.

Working: Applied to mine workings when squeezing.

Workings: The openings of a colliery, including all roads, ways, levels, dips, airways, etc.

Notes

BOOK EPIGRAPH

* The first epigraph paraphrases parts of Webber, *Thirst for Power*, 22.

PROLOGUE

1. "Spot of time": Wordsworth, *The Prelude*, 428.
2. "Historians . . . researchers": Jablonka, *History*, 2.

INTRODUCTION

1. "Keepers . . . former worlds": Morford and Lenardon, *Classical Mythology*, 130–131.

2. "Working . . . apart": Pennsylvania Department of Public Education, *Safe Practices in Mining Anthracite*, 111; "dead work": Pennsylvania Department of Public Education, *Safe Practices in Mining Anthracite*, 94.

3. "his working . . . to do": Adams, Nguyen, and Cowan, "Theories of Working Memory," 341.

4. "About . . . community": Lynch, Glotfelty, and Armbruster, *Bioregional Imagination*, 2–3; Buell, *Writing*, 246–248. The U.S. Geological Survey describes a watershed as "an area of land that drains all the streams and rainfall to a common outlet" ("Watersheds and Drainage Basins"). Others define "watershed" as "a profoundly unsettled, hybridized concept . . . existing in nested and overlapping scales, ranging from the interiority of individual bodies to the planetary hydrological cycle" (Biro, "River-Adaptiveness," 175). See Heise, *Sense of Place*, 28–49, and Buell, *Writing*, 243–265, and *Future*, 62–76, for discussions of place and bioregionalism. For "critiques of bioregionalism," see Lynch, Glotfelty, and Armbruster, *Bioregional Imagination*, 7–11.

5. For examples of the many and various histories of the Anthracite Region, see Aurand, Bodnar, Dublin and Licht, B. Folsom, Miller and Sharpless, A. Wallace, Weaver, and Wolensky and Hastie.

6. "for every . . . water": Wolensky, Wolensky, and Wolensky, *Knox*, 54. Meaning connection or link, "nexus" comes from the Latin *nectere*, to bind (*Merriam-Webster's*). Energy is "the capacity of a system to do work" (Atkins, *Laws of Thermodynamics*, 18). Power is "the rate at which work is done" (B. Johnson, *Carbon Nation*, 176).

7. "anthracite jump-started. . . United States": Chandler, "Anthracite Coal," 159.

8. "railcars . . . park": Krawczeniuk, "A Record-Breaking Year."

9. For more about "displacement without leaving home," see Albrecht et al., "Solastalgia."

10. The Delaware and Hudson Canal Company early on also sold coal in Albany and New England (Lowenthal, *From the Coalfields*, 111). Its main market, however, was New York.

11. In a contemporary sacrifice zone, producing a barrel of oil from tar sands in Canada requires 120 gallons of water (Mitchell, *Carbon Democracy*, 250); tar sands production creates "400 million gallons of toxic water a day" (B. Johnson, *Carbon Nation*, 165).

12. "In many . . . wrong": Emmett and Nye, *Environmental Humanities*, 173; "In 2018 . . . Scranton": Lockwood, "College"; "In the same . . . 1980s": Pennsylvania Department of Environmental Protection Office of Active and Abandoned Mine Operations, "Powderly Creek," and Gibbons, "Long Time Burning"; "Meanwhile . . . drilling": Wilber, *Under the Surface*, 82–92; "sacrifice . . . good": Wilber, *Under the Surface*, 218.

13. "spots of time": Wordsworth, *The Prelude*, 428.

14. "multiple places of perpetual change": Chen, "Mapping Waters," 275.

15. "In 1917": Parini, *Robert Frost*, 181. References to the poem "For Once, Then, Something" are cited in text by line number. See Frost, "For Once, Then, Something."

16. "perhaps Plato's reality": Parini, *Robert Frost*, 182.

17. Definitions in this paragraph and the following one are from *Merriam-Webster's*.

18. "With no . . . back again": Webber, *Thirst for Power*, 45.

19. "The report . . . energy": U.S. Department of Energy, *Water-Energy Nexus*, v; "Fretting . . . with water": U.S. Department of Energy, *Water-Energy Nexus*, 4.

20. "several . . . growth": Schmidt, *Water*, 2; "In 2008 . . . oil": Salzman, *Drinking Water*, 249; "In 2014 . . . nexus": Schmidt, *Water*, 178; "In 2018 . . . Decade": United Nations, "International." The official name is "International Decade for Action on Water for Sustainable Development, 2018–2028" (United Nations, "International").

21. "More than . . . production": U.S. Department of Energy, *Water-Energy Nexus*, 9; "Mainly . . . power": U.S. Department of Energy, *Water-Energy Nexus*, 18.

22. "Due to . . . by 2100": Goodell, *Water Will Come*, 175; "Siphoned . . . 1960": McNeill and Engelke, *Great Acceleration*, 157; "In the . . . sump": Pearce, *When the Rivers Run Dry*, 238–240; "multiple . . . ocean": Pearce, *When the Rivers Run Dry*, 241; "In Florida . . . America": Dearen and Schneider, "When Green Isn't Good"; "toxic . . . reservoirs": James, "Water Hazard."

23. "In 2012 . . . basin": Hauter, *Frackopoly*, 165; "governor . . . oil industry": Hauter, *Frackopoly*, 169; "Seeking . . . Apalachicola": Fineout and Sherman, "Fla., Ga. Water Fight Awaits Supreme Court"; "In 2016 . . . bottles": Fuller, "Timber Company Tells." In 2020, a judge ruled against Florida (Fowler, "New Special Master"). Texas exempts oil and gas companies from water restrictions during droughts (Tabuchi and Migliozzi, "'Monster Fracks'").

24. "Fixed . . . gods": Homer, *Odyssey*, 43–46.

25. "making . . . solvent": Strang, *Water*, 14–15; "making . . . generation": U.S. Department of Energy, *Water-Energy Nexus*, 7; "Water . . . gas": G. White, foreword to *Water in Crisis*, Pearce, *When the Rivers Run Dry*, 307, and Goodell, *Water Will Come*, 58; "Both . . . culture": Tvedt and Oestigaard, "History of the Ideas of Water," 4.

26. "date . . . fresh water": Crutzen and Stoermer, "The Anthropocene," 17–18; "Other signs . . . worldwide": Ellis, *Anthropocene*, 51; "and the . . . World War II": Ellis, *Anthropocene*, 145. Scholars have proposed several terms to describe this era, among them Anthropocene, Capitalocene, Chthulucene, and Plantationocene (Haraway, *Staying with the Trouble*, 100–101).

27. "In 2018 . . . felt": Borenstein, "'Life or Death,'" A6; "In 2019 . . . food production": Borenstein and Keaten, "Waste Land"; "In 2021 . . . events": Intergovernmental Panel on Climate Change, "Summary," 25.

28. "water vapor . . . rock": Webber, *Thirst for Power*, 44; "climate . . . problem": Morton, *Dark Ecology*, 36–37.

29. "Scientists . . . million years": Smil, *Energy*, 145, and see Finkel, *Pipeline Politics*, 172; "In 2018 . . . levels": Borenstein, "'Life or Death,'" A6; "scientists announced . . . water phenomenon": Borenstein, "Arctic"; "In 2019 . . . in 2012": Borenstein, "Hey"; "ice-core . . . decades": Maslin, *Climate Change*, 98. At the South Pole, the rate of melting has about tripled: "From 1992 to 2011, Antarctica lost nearly 84 billion tons of ice a year. From 2012 to 2017, the melt rate increased to more than 241 billion tons a year, according to . . . the journal Nature" (Borenstein, "Melting"). In 2020, major climate monitoring groups pointed out that "the six warmest years on record have been the six years since 2015" (Borenstein, "2020").

30. "The world . . . animal life": Mitchell, *Carbon Democracy*, 15; "Half . . . after 1986": Malm, *Fossil Capital*, 328; "one-quarter . . . 1999–2014": Malm, *Fossil Capital*, 353; "By 2015 . . . by 2100": Smil, *Energy*, 143.

31. "Climate . . . towns": Borenstein and Jordans, "It's a Summer of Extremes"; "By mid-August . . . average": Borenstein, "More Warmth"; "Warmer-than-usual . . . electricity": Sengupta, "The Message"; "The 2017–2018 . . . than normal": Borenstein, "The Big Chill?" Globally, over the last 35 years, the acreage of fire-ravaged land has doubled due to hotter and drier weather (Borenstein, "More Warmth"); "The summer . . . recorded": NASA, "NASA Announces."

32. "industrialization . . . catastrophes": Lieberman and Gordon, *Climate Change in Human History*, 134.

33. "In 1822 . . . quarry": Hollister, *History of the Lackawanna Valley*, 348; "In 1959 . . . field": Wolensky, Wolensky, and Wolensky, *Knox*, 125.

34. "high . . . locals": "Old Gen. Humidity on Job," 7; "Nearly . . . 4": "Excessive Heat."

35. "cratered . . . problems": Miller and Sharpless, *Kingdom of Coal*, 323–324. For a definition of peak oil, see Manno and Balogh, "The Biophysical," 43. For definitions of peak water, see Gleick and Palaniappan, "Peak Water."

CHAPTER 1

1. "During . . . earlier": Krawczeniuk, "Candidates"; "a city . . . Pittsburgh": Olson, "Report"; "When . . . coal": Micek, "Analysis"; "included . . . in Scranton": Kessler, "Biden's Claim."

2. "slow disasters": Morton, *Dark Ecology*, 69, and see Nixon, *Slow Violence*, 2; "The long . . . hardened": Smil, *Energy*, 103–104; "After . . . valley": Harrison, Marshak, and McBride, "The Lackawanna Synclinorium," 1510. Geologically, the Lackawanna and Wyoming Valleys are one formation. The Lackawanna River flows through the former, the Susquehanna River through the latter. Their confluence marks the divide between the "valleys."

3. "First . . . again": *Merriam-Webster's*; "About 750 . . . beings above": Rosen, "Snowball Earth"; Keller et al., "Neoproterozoic Glacial Origin of the Great Unconformity," 1136–1137, 1139.

4. Environmental historians early on paid attention to water. See Reisner, *Cadillac Desert*; Worster, *Rivers of Empire*; Steinberg, *Nature Incorporated*; Pisani, *To Reclaim a Divided West*.

5. A vast body of environmental writing exists. For a sampling of genres, see McKibben, *American Earth*, and Finch and Elder, *Norton Book*.

6. "As the ice . . . gravel": "Archbald's"; "waterfall . . . wide": Reese, "Outstanding."

7. "A Freak."

8. "ice . . . region": "Archbald's."

9. "curious . . . reveal it": "Archbald's"; "The Smithsonian . . . work of nature": "Archbald's"; "The site . . . falling in": "The Pot Holes Secured."

10. "Famous Freak."

11. "Two deeds . . . surrounding land": "Glacial Pothole"; "In 1967 . . . the pothole": "State, County Officials."

12. "The Wisconsin . . . gravel": Ash, *Buried Valley*, 7.

13. "Between . . . dozen": Ash, *Buried Valley*, 13.

14. "In the first . . . building": "Glacial Pot-Holes." See Stranahan, *Susquehanna*, 167n10.

15. "In December . . . pothole": "Glacial Pot-Holes"; "A torrent . . . gangways": "The Mine Horror"; "In less . . . 2,000 feet": Ash, *Buried Valley*, 14; "Attempts . . . surface": "'No Hope'"; "Early reports . . . recovered": "All Hope." Later sources confirm that 26 men died (see Ash, *Buried Valley*, 13, "Glacial Pot-Holes," and "Anthracite Mining").

16. "some places . . . workings": Ash, *Buried Valley*, 16; "Mine subsidence . . . a swamp": Ash, *Buried Valley*, 17–18; "Thirty years . . . tunnels": Ash, *Buried Valley*, 13; "To stop . . . the brook": Ash, *Buried Valley*, 16.

17. "An Algonquin . . . rock": McGurl, *Lackawanna River Guide*, 1–3; "Dropping . . . Susquehanna": Hollowell and Koester, *Ground-Water Resources*, 6; "The confluence . . . plateaus": McGurl, *Lackawanna River Guide*, 2; "In addition . . . volume": McGurl, *Lackawanna River Guide*, 19.

18. "working-class river": Lyon, "River Projects," 3. More famous working rivers include the Rhine and the Watarase (McNeill, *Something*, 131–135).

19. "In the . . . bulbs": Kashmer, "River Recollections," 10; "filthy . . . debris": Kashmer, "River Recollections," 8; "Getting . . . unmistakable": Rudzinski, "River Recollections," 13; "A former . . . live by": Kowalski, "River Recollections," 13; "Despite . . . efforts": Kashmer, "River Recollections," 10.

20. References to the poem "Lackawanna" are cited in-text by line number. See Merwin, "Lackawanna." "Part . . . west": E. Folsom, "'I Have Been,'" 242; Hix, *Understanding*, 78, 82. See Merwin, "'Fact,'" 328–329.

21. "place . . . national": Merwin, "'Fact,'" 328.

22. "Realizing . . . sees": see Hix, *Understanding*, 100.

23. "Dumping . . . paths": Leighton, *Quality of Water*, 31; "in 1916 . . . openings": Water Supply Commission of Pennsylvania, *Water Resources*, 20; "river . . . fluid": Leighton, *Quality of Water*, 24; "but it . . . yellow": Water Supply Commission of Pennsylvania, *Water Resources*, 22, 28; "Lackawanna . . . swath": Leighton, *Quality of Water*, 27.

24. "Forced . . . pollution": Freyfogle, *Land We Share*, 74; "In *Pennsylvania* . . . American law": Casner, "Acid Mine Drainage," 99; "Absolving . . . to pollute": Casner, "Acid Mine Drainage," 101.

25. "The ruling . . . mine drainage": Casner, "Acid Mine Drainage," 98–100; "and in 1923 . . . mine drainage": Tarr and Yosie, "Critical Decisions," 77–79; "it did . . . underground mines": Hughes, Hornberger, Williams, et al., "Colliery," 56; "state . . . 1965": Casner, "Acid Mine Drainage," 91.

CHAPTER 2

1. "Thomas Meredith . . . operators": Meredith, *Monopoly*, 17–18; "Maurice . . . coal trade": Lowenthal, *From the Coalfields*, 12–13, 36.

2. "Before . . . business": Yewlett, "Early Modern," 23; "Thomas's father. . . miles wide": Dickenson, "Brigadier-General," 561; "By 1812 . . . Township": Grundfest, "George Clymer," 257; "where he . . . Belmont": Dickenson, "Brigadier-General," 561; "he died . . . $13,850": Maxey, "Of Castles," 417–418, and Read, Letter; "In . . . $273,890": Official Data Foundation, Alioth Finance, "$13,850." Other sources assert that he moved to Belmont earlier. Barbe and Reed, for example, claim that Meredith moved there in 1812 (*History of Wayne County*, 220). For more information about Samuel Meredith, see "A Remarkable Man."

3. "Thomas traveled . . . Philadelphia bar": "An Illustrious Family"; "he trekked . . . life's work": Grundfest, "George Clymer," 257n69, 258; "he confessed . . . defence": Maxey, "Of Castles," 418.

4. "North and South . . . Delaware": Barbe and Reed, *History of Wayne County*, 266; "Newburgh Turnpike . . . on the Hudson": Blackman, *History of Susquehanna County*, 510; "the Corners . . . Susquehanna County": Meylert, Letter, 12.

5. "When a . . . upstate anthracite": Hudson Coal Company, *The Story of Anthracite*, 40; "skyrocketing . . . 300 percent": H. Powell, *Philadelphia's*, 146. As early as 1804, Samuel Preston, an agent for Henry Drinker, Meredith's sometime partner, knew about anthracite near Griswold's Gap (Hollister, *History of the Lackawanna Valley*, 346).

6. "Through . . . coal road": Mathews, *History of Wayne, Pike and Monroe Counties*, 576; "In 1817 . . . Stockport": "Stockport"; "The first . . . in 1820": Barbe and Reed, *History of Wayne County*, 262.

7. "in 1818 . . . the Lackawanna": *List of Lands*, and see Stocker, *Centennial History of Susquehanna County*, map, 32–33; "By 1827 . . . four feet thick": Ritter, "Art. XVII," 301. Decades later, Phineas Goodrich mentioned to state geologist I. C. White that he had long ago seen coal along the stream in Griswold's Gap (I. C. White, *Second Geological Survey of Pennsylvania*, 178). In 1830, *Hazard's Register of Pennsylvania* reported that "the Anthracite coal formation commences near the sources of the Lackawannock, not far from Belmont, the residence of Thomas Meredith, Esq" (Chapman, "Description of Luzerne County," 97).

8. "And raising . . . anyone": H. Powell, *Philadelphia's*, 88.

9. "Maurice Wurts . . . War of 1812": Lowenthal, *From the Coalfields*, 8; "they knew . . . hear Wilkes-Barre": Hollister, *History of the Lackawanna Valley*, 340; "When the . . . an opportunity": Hudson Coal Company, *The Story of Anthracite*, 40–41.

10. "While Meredith . . . Carbondale": Hollister, *History of the Lackawanna Valley*, 297; "in 1815 . . . Scranton": Ruth, *Of Pulleys and Ropes and Gear*, 4; "Wurts . . . land": Hollister, *History of the Lackawanna Valley*, 340, 341.

11. "In 1815 . . . the stones": Hudson Coal Company, *The Story of Anthracite*, 41; Ruth, *Of Pulleys and Ropes and Gear*, 4; "Still . . . Philadelphia": Lowenthal, *From the Coalfields*, 9–10, 13.

12. "investors . . . coal trade": H. Powell, *Philadelphia's*, 72; "despite . . . prices": H. Powell, *Philadelphia's*, 66; "Meanwhile . . . freshets": Hudson Coal Company, *The Story of Anthracite*, 56, and Jones, *Routes of Power*, 39–40; "In 1820 . . . the city": Hudson Coal Company, *The Story of Anthracite*, 57.

13. Ruth, *Of Pulleys and Ropes and Gear*, 5.

14. "Early on . . . Lackawanna": Lowenthal, *From the Coalfields*, 9; "After . . . sizes": H. Powell, *Philadelphia's*, 100; "At . . . winter": H. Powell, *Philadelphia's*, 94; "they . . . thick": Silliman, "Notice of the Anthracite Region," 74; "In the 1823–1824 . . . 100 tons": Hollister, *History of the Lackawanna Valley*, 348. Hollister claims that the miners worked until late fall (*History of the Lackawanna Valley*, 348). Carbondale was founded officially in April 1831 (Hollister, *History of the Lackawanna Valley*, 310).

15. "At first . . . channel": T. Murphy, *Jubilee History*, vol. 1, 135; "To keep . . . coal fire": Hollister, *History of the Lackawanna Valley*, 348. Lowenthal reads this description as more myth than fact (*From the Coalfields*, 10).

16. "Completion . . . 90 percent": Sheriff, "'Not the True Centennial,'" 373; "To reach . . . hard coal": Hudson Coal Company, *The Story of Anthracite*, 40, and H. Powell, *Philadelphia's*, 28; "followed . . . New York": Delaware and Hudson Canal Company, *Annual Report . . . 1825*, 5; "ditch . . . drought": Hudson Coal Company, *The Story of Anthracite*, 81; "opening . . . in December": Sanderson, *Delaware & Hudson Canalway*, 25–26.

17. "In 1823 . . . ends": "An Act to Improve" and Lowenthal, *From the Coalfields*, 11–12; "Pennsylvania . . . whatsoever": "An Act to Improve," 75; "As . . . *trade*": Meredith, *Monopoly*, 11.

18. "Two years . . . organized": Lowenthal, *From the Coalfields*, 26; "At about . . . Lackawaxen": "A Supplement" and Schultz and Hollister, "The Delaware and Hudson Canal Company," 121; "In 1825 . . . New York": "An Act to Amend"; "investors . . . Hone": Schultz and Hollister, "The Delaware and Hudson Canal Company," 120; "In July . . . began": Lowenthal, *From the Coalfields*, 53; "Vertically integrated": Schultz and Hollister, "The Delaware and Hudson Canal Company," 111.

19. Meredith, Letter.

20. "Jacob Cist . . . Finger Lakes": H. Powell, *Philadelphia's*, 117, 121–122.

21. "At public . . . Erie Canal": "Internal Improvement," 11 Nov. 1825; "Attendees . . . Susquehanna": "Internal Improvement," 23 Dec. 1825; "Meredith . . . intended": "Internal Improvement," 23 Dec. 1825, and Meredith, *Monopoly*, 17; "The success . . . railroads": Healey, *Pennsylvania Anthracite Coal Industry*, 57.

22. "In mid-January . . . March": "Internal Improvement," 13 Jan. 1826; "Stipulating . . . tract": *Laws of Pennsylvania*; "but also . . . coal trade": "Wayne County," 139; see Ritter, "Art. XVII," 302; "In April . . . canal": "A Further Supplement."

23. "After . . . setbacks": Schultz and Hollister, "The Delaware and Hudson Canal Company," 126; "the D & H . . . snapping": Schultz and Hollister, "The Delaware and Hudson Canal Company," 129; "and poor-quality . . . reputation": Schultz and Hollister, "The Delaware and Hudson Canal Company," 131.

24. Goodrich, *History of Wayne County*, 355–356.

25. "To adhere . . . to pass": Lowenthal, *From the Coalfields*, 63.

26. "In 1825 . . . feet": Barbe and Reed, *History of Wayne County*, 45; "In 1830 . . . injury": Meredith, *Monopoly*, 26. That $27,000 is equivalent to $851,814 today (Official Data Foundation, Alioth Finance, "$27,000").

27. "In April . . . dam": Lowenthal, *From the Coalfields*, 284n25; "After . . . own hands": Apollo, D & H Dam Destroyed.

28. "Maurice Wurts . . . the dam": Lowenthal, *From the Coalfields*, 94; "This agreement . . . grievances": Meredith, *Monopoly*, 14–15, Lowenthal, *From the Coalfields*, 95, and "Early Railroad"; "the company . . . coal boats": Lowenthal, *From the Coalfields*, 94. Larry Lowenthal points out that "as they had taken advantage of the D & H's lapse in marketing inferior coal, the anthracite rivals rejoiced in—possibly even abetted—a controversy that arose between the company and lumber raftsmen" (*From the Coalfields*, 92).

29. "engineers . . . downside levels": Schultz and Hollister, "The Delaware and Hudson Canal Company," 129; "The location . . . the route": Hollister, *History of the Lackawanna Valley*, 351; "Panther Falls . . . engines": Trails Conservation Corporation, *Upper Lackawanna Watershed Management Plan*, 32.

30. "In summer . . . canal boats": Brown, *History of the First Locomotive in America*, 75–76; "When a crew . . . embarrassment": State, "The Truth behind the 'Pride of Newcastle,'" 82. Fourteen or fifteen years after its arrival in the United States, the *Stourbridge Lion* was cannibalized for parts and the boiler taken to Carbondale for use with a stationary engine at a foundry (Brown, *History of the First Locomotive in America*, 89–90).

31. "bustling construction site": Lowenthal, *From the Coalfields*, 75.

32. Sanderson, *Delaware & Hudson Canalway*, 39.

33. "The Lion . . . too big": Lowenthal, *From the Coalfields*, 78; "With . . . horses": Schultz and Hollister, "The Delaware and Hudson Canal Company," 129; "It also . . . stocks": Ruth, *Of Pulleys and Ropes and Gear*, 26. For a description of the *Stourbridge Lion* route, see Hudson Coal Company, *The Story of Anthracite*, 87.

34. "When one . . . the river": Sanderson, *Delaware & Hudson Canalway*, 130; "and tore . . . amputated": Brown, *History of the First Locomotive in America*, 81.

35. "million-dollar": Schultz and Hollister, "The Delaware and Hudson Canal Company," 112. For a description of the gravity railroad, see Lowenthal, *From the Coalfields*, 74.

36. "Access . . . the D & H": Mathews, *History of Wayne, Pike and Monroe Counties*, 242. Philip Hone kept the company afloat (Mathews, *History of Wayne, Pike and Monroe Counties*, 242). "On October 9 . . . the canal": Hollister, *History of the Lackawanna Valley*, 300; "D & H president . . . for shipment": Meredith, *Monopoly*, 15–16; "he began . . . Dyberry": "Teams Wanted"; "In mid-November . . . this season": Meredith, *Monopoly*, 16, and see Hollister, "Samuel"; "the Wurts ditch . . . New York": Hudson Coal Company, *The Story of Anthracite*, 61. A company history states, "The primary purpose of the proposed canal was to secure a means of transporting anthracite from the Northern Wyoming Field to tidewater at New York" (Hudson Coal Company, *The Story of Anthracite*, 61). Unlike the D & H Canal, the Schuylkill Canal was not originally built to carry mainly coal; soon, however, it did (Hudson Coal Company, *The Story of Anthracite*, 49).

37. "building . . . private profit": Schultz and Hollister, "The Delaware and Hudson Canal Company," 119; "Meredith accused . . . Lackawanna Valley": Meredith, *Monopoly*, 5–7; "founding . . . anthracite": see "An Act to Improve"; "state approval . . . *HIGHWAY*": Meredith, *Monopoly*, 10, and "A Further Supplement." *Monopoly Is Tyranny; or An Appeal to the People and Legislature, from the Oppression of the Delaware & Hudson Canal Company* was first published serially in the *Wyoming Herald*, Jan. 1, 1830 (p. 2), Jan. 8, 1830 (p. 2–3), Jan. 15, 1830 (p. 2), and Jan. 22, 1830 (p. 2), the *Susquehanna Democrat*, Jan. 1, 1830 (p. 3) and Jan. 15, 1830 (p. 2), and the *Susquehanna Register*. Meredith didn't sign his name to the pamphlet, but internal evidence, the timing and place of publication, and his known opposition to the company make it likely that he wrote it. Others agree; for example, see Lowenthal, *From the Coalfields*, 96.

38. "Meredith . . . canal basin": Meredith, *Monopoly*, 16–17; "In response . . . favorable rates": Lowenthal, *From the Coalfields*, 117; "not . . . coal": S. Powell, "Chronology," 4; "By keeping . . . the Delaware": Meredith, *Monopoly*, 25.

39. "vertically integrated corporation": Schultz and Hollister, "The Delaware and Hudson Canal Company," 111; "For a decade . . . the valley": Hollister, "Samuel," 207.

40. "Shall we . . . prevent it": Lowenthal, *From the Coalfields*, 94. Meetings were held in Carbondale and Montrose on February 4 and 6, 1830, to draft resolutions in support of the D & H ("Public Meetings").

41. "Supreme Court . . . development": Kelo et al. v. City of New London.

42. "Sheriff's Sales."

43. "in 1870 . . . the river": Lowenthal, *From the Coalfields*, 234; "Meredith's proposed . . . Meredith land": "Forest City Coal"; "In 1872 . . . Susquehanna Railroad": Shaughnessy, *Delaware & Hudson*, 65. In the archives of the Lackawanna Historical Society, an 1869 map of state railroads shows a "Proposed R. R." that follows a similar route.

44. "In 1819 . . . Water Gap": Hollister, *History of the Lackawanna Valley*, 380; "Surveyed . . . of the way": Hollister, *History of the Lackawanna Valley*, 253; "Supreme Court . . . monopoly": E. Roberts, "A History of Land Subsidence," 217, and "The Judicial History of the Anthracite Monopoly," 441. Although the north-south portion of the D. L. W. has been called the Meredith Railroad (Hollister, *History of the Lackawanna Valley*, 382), the D. L. W. line ran outside of the valley, not through its bed, as Meredith projected. In 1900, J. P. Morgan controlled the six largest anthracite railroads, creating a "virtual monopoly directed from New York" (Wolensky and Hastie, *Anthracite Labor Wars*, 2; see Miller and Sharpless, *Kingdom of Coal*, 286).

CHAPTER 3

1. "In the . . . ogre": Sturluson, *Prose Edda*, 32–33; "From . . . made": Sturluson, *Prose Edda*, 35.

2. "When Odin . . . exchange": Sturluson, *Prose Edda*, 42–43; "After . . . eyes": Gardner, *On Moral Fiction*, 3.

3. "In January . . . level": "Workmen"; "Settling . . . railroad": "A Builder." The wheel pulling cars from the mine was 100-horsepower. Whenever the reservoir atop Moosic Mountain could not supply the wheel, a steam engine took over ("Extracts").

4. "To control . . . ditch": Audubon Florida, "Kissimmee"; "Crews . . . aquifers": Barnett, *Rain*, 233; "in 1992": Audubon Florida, "Kissimmee"; "crews began . . . wetlands": Barnett, *Rain*, 242.

5. "make water, or natural seepage": Wolensky and Hastie, *Anthracite Labor Wars*, 177.

6. "To cover . . . more water": Freese, *Coal*, 55; "today . . . locally": Pennsylvania Department of Environmental Protection, "Climate Change in PA."

7. "In 1917 . . . National mine": "Cave in"; "Near Forest City . . . flooding": Water Supply Commission of Pennsylvania, *Water Resources Inventory Report*, 21; "Other . . . shafts": Norris, "Unwatering," 160; "in 1913 . . . Lackawanna": Water Supply Commission of Pennsylvania, *Water Resources Inventory Report*, 23; "Maps . . . rains": McGurl, "Lackawanna River: Watershed Tour."

8. "market forces . . . parts": Miller and Sharpless, *Kingdom of Coal*, 287; "a flight . . . in the 1950s": Miller and Sharpless, *Kingdom of Coal*, 323.

9. "In the early . . . to waste": Norris, "Unwatering," 160; "Adding . . . depths": Ash, Eaton, et al., "Data on Pumping," 23; "kept . . . emergencies": Hudson Coal Company, *The Story of Anthracite*, 170; "Installing . . . be mined": "Turbine Mine." In 1931, some mines in the northern field had emergency capacity to draw "50 or more tons of water for every ton of coal produced" (Stevenson, *Reflections of an Anthracite Engineer*, 27).

10. "hand . . . capacity": Stevenson, *Reflections of an Anthracite Engineer*, 27; "In 1899 . . . a minute": "Eastern Coal and Coke Notes"; "By 1932 . . . per day": Hudson Coal Company, *The Story of Anthracite*, 155. Cut from solid rock and separated from the mine's workings, the pump room lay 270 feet beneath the surface (Hudson Coal Company, *The Story of Anthracite*, 155).

11. "In 1905 . . . in depth": Norris, "Unwatering," 159; "In 1900–1920 . . . ton of coal": Norris, "Unwatering," 160, and Wolensky, Wolensky, and Wolensky, *Knox*, 54; "In 1931 . . . 12:1": Hudson Coal Company, *The Story of Anthracite*, 155; "just six . . . 18:1": "Mining of Anthracite," 21; "Although . . . 19:1": Ash, Eaton, et al., "Data on Pumping," 22; "in the northern . . . ton of coal": Ash, Eaton, et al., "Data on Pumping," 20; "in 1951 . . . 32:1": Ash, Dierks, and Miller, *Mine Flood Prevention*, 80; "and soon . . . $12 million": Wolensky, Wolensky, and Wolensky, *Knox*, 54; "In 1957 . . . 56:1": Korb, "The Conowingo Tunnel," 1; "Without . . . collapse": Ash, Dierks, and Miller, *Mine Flood Prevention*, 82. Five hundred thirty-eight gallons of water weighs about two tons (Con version-website.com, "Tons"). One observer claimed that by 1911, mine operators in the Lackawanna Valley pumped 12–16 tons of water per ton of coal ("Turbine Mine Pumps").

12. "Defining . . . problem": Atkins, *Laws of Thermodynamics*, 38; "Discovered . . . engines": Hunt, *Pursuing Power and Light*, 26; "the first two . . . turn": Hunt, *Pursuing Power and Light*, 41; "the system . . . heat death": Hunt, *Pursuing Power and Light*, 25; "Suggesting . . . exhaustion": Hunt, *Pursuing Power and Light*, 42; "the second . . . progress": Hunt, *Pursuing Power and Light*, 25.

13. "In 1712 . . . in England": Lieberman and Gordon, *Climate Change in Human History*, 136, and Freese, *Coal*, 59, 60; "In the 1760s . . . design": Smil, *Energy*, 106; "in 1776 . . . foundry": Freese, *Coal*, 64; "a steam-powered . . . workings": Hunt, *Pursuing Power and Light*, 4–5.

14. "Just as . . . to condenser": Hunt, *Pursuing Power and Light*, 20.

15. "In 1842 . . . Cornwall": Reynolds, *Stronger Than a Hundred Men*, 275; "Beginning . . . until 1929": Reynolds, *Stronger Than a Hundred Men*, 315–316; "Watt . . . engines": Malm, *Fossil Capital*, 56; "in 1827 . . . Scotland": Malm, *Fossil Capital*, 101.

16. "In 1837 . . . company coal": Delaware and Hudson Canal Company, *Annual Report . . . 1837*, 5; "For 30 . . . planes": S. Powell, *Delaware and Hudson Canal Company*, 10, and see 11.

17. "In 1841 . . . smelter": Perry, *"A Fine Substantial Piece of Masonry,"* 10, 11; "By 1844 . . . factory": Perry, *"A Fine Substantial Piece of Masonry,"* 17; "Only in 1847 . . . replaced it": Perry, *"A Fine Substantial Piece of Masonry,"* 21, 26.

18. "In the U.S . . . 1870": Malm, *Fossil Capital*, 316; "As late . . . enterprises": Steinberg, *Nature Incorporated*, 23n5; "A ballooning . . . for water": Malm, *Fossil Capital*, 318–319; "Sealing . . . industries": Steinberg, *Nature Incorporated*, 245. In the Schuylkill Valley in 1840, a dozen steam engines worked at mining coal; by 1850, 165; by 1865, 320 (Jones, *Routes of Power*, 67).

19. "To hold . . . turbines": "Turbine Mine Pumps"; "Engineers . . . shafts": Norris, "Unwatering," 174; "A vacuum . . . each hour": Tonge, "Modern Methods," 468; "Hoisting . . . costs": Norris, "Unwatering," 177.

20. "A typical . . . lift": Norris, "Unwatering," 170; "Using . . . corrode": Norris, "Unwatering," 162; "Some . . . overnight": Norris, "Unwatering," 167; "Unlike . . . corrosion": Casner, "Acid Mine Drainage," 96.

21. "To enhance . . . pumps": "Turbine Mine Pumps"; "submersible . . . two years": "Turbine Pumps," 687; "In 1907 . . . Scranton mine": Norris, "Unwatering," 173.

22. "Corrosion . . . hours": Ash, Dierks, and Miller, *Mine Flood Prevention*, 41; "In operations . . . chutes": Ash, Dierks, and Miller, *Mine Flood Prevention*, 44; "In 1904 . . . ash": Leighton, *Quality of Water*, 24–25; "made . . . streams": Casner, "Acid Mine Drainage," 96; "in 1932 . . . washeries": Ash, Dierks, and Miller, *Mine Flood Prevention*, 44; "Chemical . . . water": Casner, "Acid Mine Drainage," 96.

23. "Completed in 1895": Corlsen, *Buried Black Treasure*, 37; "Jeddo . . . precipitation": Hughes, Hornberger, Loop, et al., "Geology," 43; "About 9,000 . . . 12 miles": Norris, "Unwatering," 168–169; "Saving . . . River": "Personal and Pertinent"; "To reduce . . . Susquehanna": Corlsen, *Buried Black Treasure*, 27. In total, operators in the eastern middle anthracite field drove 13 drainage tunnels (Hughes, Hornberger, Loop, et al., "Geology," 41).

24. "property owners . . . remining them": Korb, "The Conowingo Tunnel," 21. In 2020, the nonprofit Chesapeake Bay Foundation joined the state of Maryland in suing the U.S. Environmental Protection Agency to force Pennsylvania to do more to clean up its section of the Susquehanna watershed (Dance, "Chesapeake Bay Foundation to Sue EPA"). Agricultural runoff ranks first in polluting the bay, but ranking second is acid from abandoned mines.

25. "the tunnel . . . drainage": Korb, "The Conowingo Tunnel," 8.

26. "In 1941 . . . Pennsylvania": Ash, Dierks, and Miller, *Mine Flood Prevention*, 4; "asserted . . . problem": Ash, *Buried Valley*, 1; "A decade . . . Maryland": Korb, "The Conowingo Tunnel," 5; "The gravity-fed . . . Susquehanna": Korb, "The Conowingo Tunnel," 7; "billions . . . flooding": Weissert, "Big Oil Seeking Protection." Lateral tunnels would have drained all but the eastern middle field. Measuring 9 feet in diameter in Throop and 16 feet at Conowingo, the main tunnel would have been able to discharge 381,000 gallons of acid per minute. Estimated cost of the project in 1961: $350 million (Dierks et al., *Mine Water Control Program*, 7).

27. "High . . . gas": Ash, Dierks, and Miller, *Mine Flood Prevention*, 10; "Noting . . . domestic coal": Ash, Dierks, and Miller, *Mine Flood Prevention*, 10, 27; "In 1957 . . . to 100": Ash, Dierks, and Miller, *Mine Flood Prevention*, 29; "The southern . . . 420 years": Ash, Dierks, and Miller, *Mine Flood Prevention*, 31; "Across . . . centuries": Ash, Dierks, and Miller, *Mine Flood Prevention*, 10.

28. "The field's . . . Rivers": Ash, Dierks, and Miller, *Mine Flood Prevention*, 35; "In 1948 . . . runoff": Ash, Dierks, and Miller, *Mine Flood Prevention*, 23; "Of the 45 billion . . . Lackawanna River": Ash, Dierks, and Miller, *Mine Flood Prevention*, 23, and see Hollowell and Koester, *Ground-Water Resources of Lackawanna County*, 39, 41–42; "Wyoming Valley . . . Lackawanna mines": Ash, Dierks, and Miller, *Mine Flood Prevention*, 24; "In total . . . Susquehanna": Ash, Dierks, and Miller, *Mine Flood Prevention*, 24; "Within . . . 23": Ash, Dierks, and Miller, *Mine Flood Prevention*, 36; "it stood . . . issues": Ash, Dierks, and Miller, *Mine Flood Prevention*, 6.

29. References to the poem "The Well" are cited in text by line number. See Merwin, "The Well."

30. Graves, *The Greek Myths*, 179–180.

31. "The Greeks . . . Tartarus": Strang, *Water*, 19.

32. "in the 1940s . . . gallons yearly": McGurl, Hughes, and Hewitt, *Lower Lackawanna River*, 51; "When pumps . . . Cayuga Lake": EPCAMR, "3D Mine Pool Mapping."

33. "expands . . . subsidence": Appalachian Regional Commission, Department of Environmental Protection, *Relationship between Underground Mine Water Pools*, x–xi, 82, and Hollowell and Koester, *Ground-Water Resources of Lackawanna County*, 36; "A typical . . . 24 hours": McGurl, "Lackawanna River: Watershed Tour"; "Acid . . . in fall": Hughes, Hornberger, and Hewitt, "The Development and Demise of Major Mining," 81.

34. "The pools . . . metals": Hughes and Hornberger, "Introduction to the Anthracite Coal Region," 18; "Surface waters . . . mix": "Dimictic Lake."

35. "as collieries . . . more mines": Ash, Dierks, and Miller, *Mine Flood Prevention*, 26, and Hollowell and Koester, *Ground-Water Resources of Lackawanna County*, 36; "By 1957 . . . *impossibility*": Ash, Dierks, and Miller, *Mine Flood Prevention*, 27. For amounts of rainfall recorded in Scranton and amounts of water pumped from mines in 1944–1951 in the area that contributes to the Scranton mine pool, see Hollowell and Koester, *Ground-Water Resources of Lackawanna County*, 36–37.

36. "If not . . . the latter": Ash, Dierks, and Miller, *Mine Flood Prevention*, 29; "Already . . . futures": Ash, Dierks, and Miller, *Mine Flood Prevention*, 90; "To keep . . . the total": Ash, Dierks, and Miller, *Mine Flood Prevention*, 31.

37. "To prevent . . . between workings": Wolensky, Wolensky, and Wolensky, *Knox*, 66; "By 1952 . . . 150 pools": Ash, Dierks, and Miller, *Mine Flood Prevention*, 26; "Lake Wallenpaupack . . . gallons": Ash, Dierks, and Miller, *Mine Flood Prevention*, 24; "In the Wyoming . . . 48 mines": Ash, Dierks, and Miller, *Mine Flood Prevention*, 28–29; "If Lackawanna . . . mining": Ash, Dierks, and Miller, *Mine Flood Prevention*, 90.

38. "Early . . . pillars": Ash, Dierks, and Miller, *Mine Flood Prevention*, 28; "nor did . . . Susquehanna": Ash, Dierks, and Miller, *Mine Flood Prevention*, 16; "As early . . . Hazleton": Stepenoff, *Their Fathers' Daughters*, 81; "by 1951 . . . mined": Ash, Dierks, and Miller, *Mine Flood Prevention*, 36; "By about . . . production": Hughes and Hornberger, "Introduction to the Anthracite Coal Region," 11; "Hundreds . . . uncertainty": Ash, Dierks, and Miller, *Mine Flood Prevention*, 26. In the anthracite industry's history, about 90 percent of mining happened deep underground (Korb, "The Conowingo Tunnel," 1).

39. For a discussion of the loss of "ocean memory . . . which may present previously unknown challenges for predicting climate extremes," see Shi et al., "Global Decline."

40. B. Johnson, *Carbon Nation*, 78.

41. "In Old Forge . . . to act": Falchek, "Old Forge," A1, and see "City Firm," 8; "in 1962 . . . Colliery": "City Firm," 3, 8. For information about the amount of the over-

flows at Old Forge and Duryea in 1962–1970, see Hollowell and Koester, *Ground-Water Resources of Lackawanna County*, 37.

42. "Gushing . . . minute": Korb, "The Conowingo Tunnel," 8; "over 130 . . . day": McGurl, *Lackawanna River Guide*, 19; "no fishery . . . survives": McGurl, *Lackawanna River Guide*, 6; "the river . . . Bay": Falchek, "Old Forge," A1.

43. "plans . . . uses": Rosler, "Light at the End of the Borehole."

44. "In 2013 . . . backyards": Gibbons and Skrapits, "DEP to Fill Mine Voids in Luzerne."

45. "planned . . . no tunnel": Ash, Dierks, and Miller, *Mine Flood Prevention*, 48; "problems . . . study": Ash, Dierks, and Miller, *Mine Flood Prevention*, 74; "treated . . . seawater": Korb, "The Conowingo Tunnel," 9; "just as . . . from the tunnel": Ash, Dierks, and Miller, *Mine Flood Prevention*, 74–75; "downplay . . . dam": Ash, Dierks, and Miller, *Mine Flood Prevention*, 91; "planners . . . wastes": Ash, Dierks, and Miller, *Mine Flood Prevention*, 74; "In the 1940s . . . 3.2": Ash, Dierks, and Miller, *Mine Flood Prevention*, 43; "Later . . . 5.3": Korb, "The Conowingo Tunnel," 10.

46. "mine water . . . Conowingo Dam": Korb, "The Conowingo Tunnel," abstract; "dry times . . . the dam": Ash, Dierks, and Miller, *Mine Flood Prevention*, 75; "Greater . . . even more": Korb, "The Conowingo Tunnel," 12.

47. "Too expensive . . . (AMWCP)": Dierks et al., *Mine Water Control Program*, 8; "AMWCP . . . broken land": Korb, "The Conowingo Tunnel," 12; "With no . . . Leuders Creek": Korb, "The Conowingo Tunnel," 15–16; "Lackawanna . . . basin": Korb, "The Conowingo Tunnel," 21–22. Total cost of the Conowingo plan in 1957: $280 million (Ash, Dierks, and Miller, *Mine Flood Prevention*, 76).

48. "In October . . . 300,000 fish": Korb, "The Conowingo Tunnel," 14; "affected drinking . . . to recover": "Fish Commission"; "The Mines . . . lose work": "Mine Drainage"; "In December . . . the job": "Glen Alden Again."

49. "The fish kill . . . Streams Act": Korb, "The Conowingo Tunnel," 14; "In 1965 . . . flooded": Hughes and Hornberger, "Introduction to the Anthracite Coal Region," 17.

50. Dierks et al., *Mine Water Control Program*, 7, 22.

CHAPTER 4

1. "Economists . . . embodied, water": Pearce, *When the Rivers Run Dry*, 5.

2. "after 1911 . . . compensation": Fishback and Kantor, "Adoption of Workers' Compensation," 318; "until 1969 . . . black lung": Wolensky, Wolensky, and Wolensky, *Fighting for the Union Label*, 145.

3. "As sales . . . cotton companies": Malm, *Fossil Capital*, 54; "When steam . . . pumped water": Lieberman and Gordon, *Climate Change in Human History*, 137; "soon powered machinery": Smil, *Energy*, 107; "first automatic device": Malm, *Fossil Capital*, 66; "cop, or quill": *Merriam-Webster's*; "In the 1780s . . . cotton mills": Hunt, *Pursuing Power and Light*, 11; "initiating . . . businesses": Lieberman and Gordon, *Climate Change in Human History*, 138; "ensuring . . . Revolution": Tvedt, "Why England," 42.

4. "before . . . Mine Workers": "Mine Workers." For examples of separate discussions of mining and textile making, see Aurand, *Coalcracker*, and Stepenoff, *Their Fathers' Daughters*.

5. "Progressive . . . cleanliness": Hoy, *Chasing Dirt*, 87–89; "At the height . . . through woods": "Bad Bathing Place"; "One August . . . a bath": "Pittston," *Scranton Tribune*.

6. "The *Scranton Tribune* . . . suburbs": "Public Baths for Scranton"; "*Scranton Times* . . . for bathing": "The Need of Public Baths"; "Paying . . . and cities": "As to Public Baths"; "local leaders . . . 1909": "Hurrah for Fourth; Big Celebration."

7. "In 1904 . . . accommodates all": P. Roberts, *Anthracite Coal Communities*, 143–145; "filth . . . the body": P. Roberts, *Anthracite Coal Communities*, 337.

8. "Pennsylvania . . . each colliery": Stevenson, *Reflections of an Anthracite Engineer*, 32–33, and see P. Roberts, *Anthracite Coal Communities*, 144; "Company houses . . . dirty": J. White, *Miner's Wash and Change Houses*, 5; "An in-between . . . and work": J. White, *Miner's Wash and Change Houses*, 21; "a wash house . . . to keep clean": J. White, *Miner's Wash and Change Houses*, 8–9; "maintained warm . . . to the ceiling": J. White, *Miner's Wash and Change Houses*, 12–14; "Although . . . in wash houses": J. White, *Miner's Wash and Change Houses*, 17, and see Sawyer, "Change House with Swimming Pools."

9. "Operators . . . workforce": Stepenoff, *Their Fathers' Daughters*, 29; "area's . . . single women": Stepenoff, *Their Fathers' Daughters*, 27, 52–53; "In Scranton . . . 159 men": Stepenoff, *Their Fathers' Daughters*, 32; "The local . . . powered them": Chittick, *Silk Manufacturing and Its Problems*, 9. In 1890, women outnumbered men in the Sauquoit mill 600 to 35 (Hall, "If Looms Could Speak").

10. "Opened . . . mill": Stepenoff, *Their Fathers' Daughters*, 25; "each month . . . silk goods": Hall, "If Looms Could Speak"; "raw silk . . . for silk linings": "The Silk Industry," 8; "In 1928 . . . production": T. Murphy, *Jubilee History*, vol. 1, 127, and B. Folsom, *Urban Capitalists*, 115. The process for warp, or organzine, is similar, with threads twisted "twelve or more turns to the inch" ("The Silk Industry," 8).

11. "With . . . to moisture": Chittick, *Silk Manufacturing and Its Problems*, 125; "owners . . . 70°F temperature": Chittick, *Silk Manufacturing and Its Problems*, 132; "To create . . . to cool bodies": Chittick, *Silk Manufacturing and Its Problems*, 127–130; "typical . . . days": Stepenoff, *Their Fathers' Daughters*, 75.

12. "Developed in 1856": United States Tariff Commission, *Report on Dyes*, 15; "By 1918 . . . its use": United States Tariff Commission, *Report on Dyes*, 34–35.

13. "Dyeing . . . water": Chittick, *Silk Manufacturing and Its Problems*, 10; "the purer, the better": Chittick, *Silk Manufacturing and Its Problems*, 99; "An outfit . . . gallons": Chittick, *Silk Manufacturing and Its Problems*, 106; "Dye houses . . . little soot": Chittick, *Silk Manufacturing and Its Problems*, 10, 100.

14. "Most . . . mills": T. Murphy, *Jubilee History*, vol. 1, 127; "In Carbondale . . . 17 years old": Stepenoff, *Their Fathers' Daughters*, 40; "In all . . . 16": Stepenoff, *Their Fathers' Daughters*, 102; "In 1907 . . . Saturday": Stepenoff, *Their Fathers' Daughters*, 75.

15. "In 1900–1901 . . . elsewhere": "Silk Mill Hands" and "An Example"; "The local . . . children": "Mine Workers"; "special . . . attention": "Strikers Gaining Ground"; "standing . . . too much": "The Silk Mill Strike"; "beginners . . . earnings": "Silk Mills Strike"; "adults . . . self-support": "Extension of the Strike"; "In February . . . eight weeks": "The Silk Workers' Wages"; "After . . . South": "They Are Happy Now." In 1999, only 1,300 workers remained in the apparel industry in the Scranton/Wilkes-Barre area (Wolensky, Wolensky, and Wolensky, *Fighting for the Union Label*, 203). In 1990, the United States made about 50 percent of its clothing; by 2013, this had dropped to 2 percent (Cline, *Overdressed*, 5). At about the same time, Americans doubled their clothing purchases, and textile waste jumped by 40 percent (Cline, *Overdressed*, 224), mainly due to cheap imports made by workers toiling in unsafe conditions. An example: a garment

factory in Dhaka, Bangladesh, collapsed in 2013, "killing at least 1,129 workers, many of whom made clothes for Western brands" (Cline, *Overdressed*, 224).

16. "Just as . . . selling coal": "Washing Anthracite"; "Colliers showered . . . superior coal": Lauver, "A Walk through the Rise and Fall of Anthracite Might"; "Some . . . longer": Hudson Coal Company, *The Story of Anthracite*, 198; "No matter . . . gas": Miller and Sharpless, *Kingdom of Coal*, 287.

17. "The Glen Alden . . . competitors": "Blue Coal Is Here"; "Reading . . . red dye": Wolensky and Hastie, *Anthracite Labor Wars*, 92; "Hudson . . . silver": Sedlisky, "Huber: The End of an Era"; "The Pennsylvania Coal . . . pink": Pennsylvania Coal Company, *The Pennsylvania Story*, and "Blue Coal, Pink Coal"; "dyeing may . . . coal cars": Sedlisky, "Huber: The End of an Era."

18. "In the 1950–1960s . . . beige": Kashmer, "River Recollections," 11; "The river . . . clear water": Kowalski, "River Recollections," 13. Rivers elsewhere are also changing color. In China, dyes color the Pearl River red and indigo (Cline, *Overdressed*, 124). Since 1984, thousands of miles of American rivers have changed color due to fertilizers and sediments; about 5 percent are an original blue, 28 percent are green, and 66 percent are yellow (Borenstein, "Waterways").

19. "American . . . success": K. Robertson, "Oil Futures/Petrotextiles," 244; "lace-curtain, or aspiring to middle-class standing": *Merriam-Webster*'s.

20. "Looms . . . yarns": Tramontana, "Black Iron, White Lace," 8; "To discover . . . pressing": "The Lace Factory"; "The lace-making . . . white color": Tramontana, "Black Iron, White Lace," 11.

21. "Easy . . . Scranton": Tramontana, "Black Iron, White Lace," 8; "one time . . . mines": "Notes of Tour"; "Employing . . . looms": "Nottingham Lace Manufacturers"; "Early . . . lace": Haggerty, "Sew It Goes," H6; "During . . . packing": "Scranton Corp. Begins 66th Year."

22. Belin, "Address to the Scranton Chamber of Commerce."

23. "In the early . . . white yet": Taber and Taber, *The Delaware, Lackawanna, & Western Railroad*, vol. 2, 394–395.

24. Phoebe is another name for Artemis, the virgin huntress of Greek mythology (*Merriam-Webster's*). "featured . . . still bright": Taber and Taber, *The Delaware, Lackawanna, & Western Railroad*, vol. 2, 392; "Maiden . . . Snow": Taber and Taber, *The Delaware, Lackawanna, & Western Railroad*, vol. 2, 398; "Promoting . . . 80 percent": Taber and Taber, *The Delaware, Lackawanna, & Western Railroad*, vol. 2, 392.

25. "Ads in . . . Anthracite": Voss-Hubbard, "Hugh Moore and the Selling of the Paper Cup," 90, and Lackawanna Railroad, *The Story of Phoebe Snow*; "Underscoring . . . conditions": "Lackawanna Railroad."

26. "it means . . . picturesque": "The Story of Phoebe Snow"; "Each . . . anthracite": "The Road of Anthracite."

27. "She made . . . Department": Taber and Taber, *The Delaware, Lackawanna, & Western Railroad*, vol. 2, 399; "In the 1960s . . . the Lackawanna": Halpin, "Phoebe Snow Travels West for Final Time," 23; "whose . . . soft coal": "Phoebe Snow Symbolized Cleanliness" and Schechner, "Phoebe Snow's Friends Sadly Say Farewell."

28. "To wear . . . use": *Merriam-Webster's*; "clothing . . . coal": K. Robertson, "Oil Futures/Petrotextiles," 243; "To make . . . of water": Cline, *Overdressed*, 125, and see Siegle, *To Die For*, 105–106; "clothes make . . . chemicals": Conca, "Making Climate Change Fashionable"; "fleece . . . marine life": K. Robertson, "Oil Futures/Petrotextiles," 256.

CHAPTER 5

1. "The dirty . . . Pittston": "Pittston," *Scranton Tribune*; "Culm . . . River": "Culm in the River," *Wilkes-Barre Weekly Times*; "By the time . . . August heat": "Culm in the River," *Wilkes-Barre Times*; "the culm . . . benefitted it": "Culm in the River," *Wilkes-Barre Weekly Times*. The *Wilkes-Barre Times* and the *Wilkes-Barre Weekly Times* offer disputing accounts of the seriousness and size of the wastewater flow.

2. "billows . . . flags": Evans, "Reserve Our Anthracite for Our Navy," 248, 249; "Aboard . . . powder": "On the Wing"; "In 1907 . . . in the navy": Evans, "Reserve Our Anthracite for Our Navy," 246; "$18 billion": Evans, "Reserve Our Anthracite for Our Navy," 253; "fantastic scheme": "Would Seize All Anthracite."

3. "Phoebe threatens . . . Anthracite": "Phoebe Snow Replies to Admiral Evans"; "Morgan . . . mines": Miller and Sharpless, *Kingdom of Coal*, 286.

4. "rock into . . . rooms": Pennsylvania Department of Public Education, *Safe Practices in Mining Anthracite*, 98; "Breakers . . . method of washing": Hudson Coal Company, *The Story of Anthracite*, 185; "boys . . . railcars": Healey, *The Pennsylvania Anthracite Coal Industry*, 40, and Stevenson, *Reflections of an Anthracite Engineer*, 29; "In 1904 . . . washing": P. Roberts, *Anthracite Coal Communities*, 174–175.

5. "Machine . . . field": "Washing Anthracite"; "by 1895 . . . wet methods": Stevenson, *Reflections of an Anthracite Engineer*, 30; "three . . . ton of coal": Prochaska, *Coal Washing*, 339; "In the 1920s . . . Chance cone": "Chance Cone Cleaned Coal" and Hudson Coal Company, *The Story of Anthracite*, 187; "As coal . . . sizes": Hudson Coal Company, *The Story of Anthracite*, 190; "Coal is . . . sizing screens": Hudson Coal Company, *The Story of Anthracite*, 188; "To catch . . . sand": Hudson Coal Company, *The Story of Anthracite*, 190–191; "anthracite . . . appearance": "Washing Anthracite"; "Many collieries . . . too costly": Prochaska, *Coal Washing*, 340; "In the 1940s . . . washing coal": Ineson and Ferree, *Anthracite Forest Region*, 14.

6. Hudson Coal Company, *The Story of Anthracite*, 192–195. For a diagram of anthracite washing, see Foster, *The Coal and Metal Miners' Pocketbook*, (1902), 442.

7. "Schuylkill . . . the total": Ineson and Ferre, *Anthracite Forest Region*, 8; "In 1945 . . . silt": Stewart, "Stream Pollution Control in Pennsylvania," 592.

8. "operators . . . washeries": "Washing Anthracite"; "Although . . . years ago": Stevenson, *Reflections of an Anthracite Engineer*, 31; "This ongoing . . . mine-scarred land": "Ensure Funds to Continue Recovery Work."

9. "body is . . . water": U.S. Geological Survey, "The Water in You"; "about 67 percent carbon": LeCain, *The Matter of History*, 332.

10. "The modern . . . environment": Alaimo, *Bodily Natures*, 90; "Transcorporality": see Alaimo, *Bodily Natures*, 2–4.

11. "as clean . . . industry": Hudson Coal Company, *The Story of Anthracite*, 218.

12. "At the end . . . itself": Sedlak, *Water 4.0*, 45–46; "rivers do . . . dilution": Steinberg, *Nature Incorporated*, 214; "If a stream . . . clean": Steinberg, *Nature Incorporated*, 206.

13. "out of sight, out of mind": Steinberg, *Nature Incorporated*, 211; "In 1890 . . . epidemic": Sedlak, *Water 4.0*, 44; "Opening . . . percent": Sedlak, *Water 4.0*, 52; "This success . . . assumed": Steinberg, *Nature Incorporated*, 239.

14. "In 1885 . . . 1,200–1,300": Melosi, *Sanitary City*, 93; "People . . . with culm": P. Roberts, *Anthracite Coal Communities*, 334; "in 1905 . . . sickening 132": "The Typhoid Situation"; "in December . . . killed 123": "Boil Drinking Water"; "Three . . .

hospitals": "28 Typhoid Cases in Hospital." State health officials traced the cause of the Scranton outbreak to the Elmhurst watershed ("Boil Drinking Water"). In 1895, officials feared that contamination from a tannery in the watershed might cause an epidemic in the city ("Purity of City Water").

15. "Advertisers . . . river": "Blamed Their Fathers."

16. "coal operators . . . sewage": Hudson Coal Company, *The Story of Anthracite*, 156–157.

17. "The town . . . terrible": "The Board of Health"; "In 1891 . . . the Lackawanna": "In Common Council"; "allowing . . . at will": "A Big Sewer Scheme"; "A year . . . unendurable": "Our Drinking Water."

18. "The night soil . . . plague": "Dying from Neglect"; "When they . . . mine water": Casner, "Acid Mine Drainage," 98. The exclusion was due to the 1886 state supreme court ruling in *Pennsylvania Coal Company v. Sanderson* (Casner, "Acid Mine Drainage," 98–99). Scranton built a crematory by 1895 ("Purity of City Water").

19. References to the poem "The River" are cited in-text by line number. See Kelley, "The River."

20. "When the . . . contrary": "What Some People Say"; "In 1902 . . . organic matter": "Streams Polluted with Mine Refuse."

21. "In 1904 . . . the stream": Leighton, *Quality of Water*, 31; "fine coal . . . Wilkes-Barre": Leighton, *Quality of Water*, 31, and Stranahan, *Susquehanna*, 164; "unfit . . . carried": Leighton, *Quality of Water*, 23; "the river . . . purified": Leighton, *Quality of Water*, 70. In 1903, the Department of the Interior established a hydro-economics division to conduct a "systemic investigation of the quality of surface and ground waters in the United States" (Walcott, *Twenty-Fourth Annual Report*, 214). Martin Melosi points out that "not until after World War I were industrial wastes viewed as a major problem affecting water purity and sewage treatment. In some cases, industrial waste was considered a germicide that, when added to water, would inhibit the putrescibility of organic material" (*Sanitary City*, 96). In 1948, a U.S. Bureau of Mines report concluded that "sewage and industrial wastes that find their way into the receiving streams, is at present distinctly beneficial rather than detrimental" (Felegy, Johnson, and Westfield, *Acid Mine Water*, 48).

22. "Noting . . . pollutant": Ash, Dierks, and Miller, *Mine Flood Prevention*, 36–37.

23. "In the 1940s . . . harmless": Rudzinski, "River Recollections," 14; "Pointing to . . . treatment plant": "Pollution Fight Is Planned," Rudzinski, "River Recollections," 14, and McGurl, *Lackawanna River Guide*, 10; "In 1944 . . . anthracite industry": "Pollution Fight Is Planned" and Rudzinski, "River Recollections," 14; "In the 1950s . . . safe distance": Kashmer, "River Recollections," 7–8; "Responding . . . the river": Stranahan, *Susquehanna*, 164. See the August 30, 1942, *Scrantonian* for a photo of the Lackawanna in the late 1880s. The caption reads, "This is how the Lackawanna River looked back in the 1880s. It abounded in fish and was used on occasion for baptisms" ("Rotogravure Section," 89).

24. "In 1962 . . . Lackawanna River": "Health Officials Offer Council Sewage Report"; "Following up . . . illegal": "Mine Voids Use for Sewage Is Termed Illegal," 3; "Although . . . 1966": McGurl, *Lackawanna River Guide*, 10; "Scranton . . . in 1991": J. Murphy, "Four Local Sewage Plants," A1; "In 1999 . . . tunnels": Golay, "Taylor Residents to Air Demands."

25. "By 1973 . . . Valley": McGurl, et al., *Lackawanna River Watershed Conservation Plan*, 5.1; "modern sewage . . . microorganisms": Strang, *Water*, 36; "In 2012 . . . Bay": Lamm et al., *Sewage Overflows in Pennsylvania's Capital*, 9; "To protect . . . 20 years": Lockwood, "Upgrades Due to Antiquated Sewer Lines."

26. "Between . . . natural gas": Aldrich, "An Energy Transition before the Age of Oil," 7; "Anthracite . . . cleanliness": Aldrich, "An Energy Transition before the Age of Oil," 3; "All three . . . on price": Aldrich, "An Energy Transition before the Age of Oil," 12–13; "Despite . . . washeries": Aldrich, "An Energy Transition before the Age of Oil," 20; "hard coal . . . to gas": Aldrich, "An Energy Transition before the Age of Oil," 23.

27. "In 2007 . . . danger": Romm, *Climate Change*, 181; "Defining . . . production": Maslin, *Climate Change*, 129; "In 2015 . . . review": EPA, "Clean Power"; "In 2017 . . . more mining": Daly and Colvin, "'A New Era in Energy Production'"; "In 2020 . . . annually": "Environmental Vandalism"; "The next . . . force": Brady, "Biden Signs Bill." In 2022, the U.S. Supreme Court overturned the Clean Power Plan (West Virginia v. EPA).

28. "Some . . . waste": Crable, "Radioactive Remains"; "Natural gas . . . and oil": Finkel, *Pipeline Politics*, 181.

29. "Coal companies . . . voters accept": Hopton and Vernon, "From the Sublime to Destruction"; "Between 2007 and 2010": Gold, *The Boom*, 253; "few people . . . Foundation": Gold, *The Boom*, 249–250.

30. "Blaschak . . . footprint": Allabaugh, "Mining a Milestone," A21.

31. "collect . . . power plants": Lieberman and Gordon, *Climate Change in Human History*, 191; "in 2010 . . . clean energy": B. Johnson, *Carbon Nation*, 167; "In his . . . clean coal": Trump, "President Donald J. Trump's State of the Union Address." In his delivery of the address, the president added the word "beautiful" ("Fact Check"). His administration's dismissal of climate change extended to removing mention of it from websites of federal agencies, including the EPA (Davenport, "'Climate Change' Mentions Scrubbed from Fed Websites"). "In 2023 . . . success": General Accounting Office, "Carbon Capture and Storage."

32. "Although . . . systems": Allabaugh, "Mining a Milestone," A21, and Hudson Coal Company, *The Story of Anthracite*, 382; "At least . . . as sand": Farrell, "The Use of Anthracite Coal as a Filter Medium," 718, 724; "Carbon Sales . . . nearly pure": O'Connell and Allabaugh, "From NEPA, with Love," C5; "In the 1950s . . . alone": Conley, "Experience with Anthracite-Sand Filters," 1473.

33. "Over 1 . . . it": Webber, *Thirst for Power*, 11, Strang, *Water*, 147, and see World Health Organization, "1 in 3 People"; "Over two . . . per day": Yaeger, "Introduction: Dreaming of Infrastructure," 16, and World Health Organization, "1 in 3 People"; "Ballooning . . . by 2030": Maslin, *Climate Change*, 82–83.

34. "About 800 . . . coastal waters": Sedlak, *Water 4.0*, 119; "The chemicals . . . 110 million": "U.S. Drinking Water Contamination" and see Dennis, "'Not a Problem You Can Run Away From,'"; "As . . . Michigan": Salzman, *Drinking Water*, 145, and see 139–157; "pipes in . . . modern era": Sedlak, *Water 4.0*, 106. See Sedlak, *Water 4.0*, 91–92 and 153, for discussions of chemicals in water supplies.

35. "Responding . . . in 1972": Sedlak, *Water 4.0*, 87; "dumping . . . wastes": Marcus, "State Grapples"; "In 1979 . . . 60 miles": C. Robertson, "The Butler Water Tunnel"; "may . . . Harrisburg": Brubaker, *Down the Susquehanna*, 60; "The interstate . . . into mines": C. Robertson, "The Butler Water Tunnel"; "For several . . . tunnel drained":

Brubaker, *Down the Susquehanna*, 60, and C. Robertson, "The Butler Water Tunnel"; "The cleanup . . . Superfund program": C. Robertson, "The Butler Water Tunnel"; "Although . . . toxic mess": Brubaker, *Down the Susquehanna*, 60, and C. Robertson, "The Butler Water Tunnel."

36. "The day . . . for anthracite": Scholz, "State Says Man."

37. "In 2020 . . . to millions": Knickmeyer, "Trump Lifts Waterways Protections"; "The revision . . . for drinking water": "He Loves That Dirty Water"; "At the same . . . arsenic": Friedman, "E.P.A. Relaxes Rules"; "In the waning days . . . air and water": Eilperin and Dennis, "EPA Rule Requires Raw Research Data"; "Biden . . . reversals": Associated Press, "The EPA Finalizes," Boyle, "Biden Reverses Trump Rollback," and EPA, "EPA Implements"; "in 2023 . . . protections": Phillis, Daly, and Flesher, "Biden Administration."

38. "In 2014 . . . 300,000 people": Lassiter, "The Elk River Spill," 17; "Declaring . . . bath in it": Lassiter, "The Elk River Spill," 22; "Forcing . . . 100,000": Lassiter, "The Elk River Spill," 24; "one-third . . . uncommon": Lassiter, "The Elk River Spill," 17–18.

CHAPTER 6

1. "Lamenting . . . Ararat": Genesis, chap. 7–8; "God strikes . . . globe": Genesis 9:11; "After . . . subdue it": Genesis 9:7.

2. "The Yellow . . . the planet": Pearce, *When the Rivers Run Dry*, 108; "In 1887 . . . 3.7 million": Gleick, "An Introduction to Global Fresh Water Issues," 11; "to stop . . . 890,000": Pearce, *When the Rivers Run Dry*, 107–108; "In the U.S . . . .the poor": Barnett, *Rain*, 148–150.

3. "In 2016 . . . North America": Hanson and Brown, "Turbulent Air"; "In another . . . penalties": McGlashen, "Biden Administration."

4. "A flood . . . a massacre": Aurand, "The Lattimer Massacre," 6; "After a subsidence . . . drown men in Ebervale": Pennsylvania Department of Internal Affairs, *Reports of the Inspectors of Mines of the Anthracite Coal Regions of Pennsylvania, 1885*, 119–120.

5. Pennsylvania Department of Internal Affairs, *Reports of the Inspectors of Mines of the Anthracite Coal Regions of Pennsylvania, 1885*, 120–123.

6. "Millions . . . Tunnel": Pennsylvania Department of Internal Affairs, *Reports of the Inspectors of Mines of the Anthracite and Bituminous Coal Regions of Pennsylvania, 1891*, 141; "a seven-by-nine-foot . . . Mountain": Aurand, "The Lattimer Massacre," 6; "Completed . . . 1895": "Coal Mines Reclaimed"; "the tunnel . . . Lattimer Massacre": Aurand, "The Lattimer Massacre:," 7–9. For more about Lattimer, see Novak, *Guns of Lattimer*, and Shackel, *Remembering Lattimer*.

7. "To prevent . . . between workings": Pennsylvania Department of Internal Affairs, *Reports of the Inspectors of Mines of the Anthracite and Bituminous Coal Regions of Pennsylvania, 1891*, 75–77; "Although . . . legal": Wolensky and Hastie, *Anthracite Labor Wars*, 376; "the law . . . robbed them": Wolensky and Hastie, *Anthracite Labor Wars*, 176–177; "but also . . . failing pillars": Wolensky, Wolensky, and Wolensky, *Knox*, 66, 68.

8. "With few . . . abandoned": "Mines and Mining"; "In 1891 . . . ahead of them": "Two Mine Horrors."

9. "On the same . . . had drowned": "A Fearful Fate"; "but rescuers . . . didn't make it": "A Mining Miracle"; "A full . . . survivors": "A Mining Miracle" and Pennsylvania

Department of Internal Affairs, *Reports of the Inspectors of Mines of the Anthracite and Bituminous Coal Regions of Pennsylvania, 1891*, 152; "who had . . . urine": "A Mining Miracle" and "19 Days Entombed."

10. "In 1899 . . . hoisting bucket": "Serious Flood in Schooley Mine"; "The company . . . dams": "Drowing [*sic*] the Schooley"; "but 10 days . . . the flood": "Flood in the Schooley Mine" and "Water Still Coming In"; "After another . . . boiler house": "Cave after Water"; "A month . . . in defeat": "Mine Allowed to Fill Up"; "From then . . . rise": "Told in Brief." Knox Coal Company would later lease the Schooley mines, which interconnected with Knox operations at River Slope (Wolensky, Wolensky, and Wolensky, *Knox*, 8).

11. "In 1924 . . . out of work": "Lackawanna River Pours into Glen Alden Mines," 1; "men tossed . . . boxcars": "Lackawanna River Pours into Glen Alden Mines," 4; "Hours later . . . two months": "Begin Task of Removing Water."

12. "In 1891 . . . lay idle": Pennsylvania Department of Internal Affairs, *Reports of the Inspectors of Mines of the Anthracite and Bituminous Coal Regions of Pennsylvania, 1891*, 100–101; "In 1902 . . . pumping operation": "To Flood the Mine"; "In 1920 . . . costly cave-ins": "Water Thwarts Mayor from Entering Mine," 1.

13. "In 1950 . . . result": Ash, *Buried Valley*, 3.

14. "To close . . . mine cars": Wolensky, Wolensky, and Wolensky, *Knox*, 50, 52; "When a . . . undid it": Wolensky, Wolensky, and Wolensky, *Knox*, 50; "laborers . . . shavings": Wolensky, Wolensky, and Wolensky, *Knox*, 45; "On the third day . . . half mile": Wolensky, Wolensky, and Wolensky, *Knox*, 52.

15. "The dewatering . . . 142,000 per minute": Wolensky, Wolensky, and Wolensky, *Knox*, 54–55; "drawing out . . . poured in": Wolensky, Wolensky, and Wolensky, *Knox*, 52, 110; "After . . . sand": Wolensky, Wolensky, and Wolensky, *Knox*, 55–58.

16. "Although . . . rising river": "12 Missing and 35 Rescued," 2; "formal investigation . . . buried valley": Wolensky and Hastie, *Anthracite Labor Wars*, 179; "companies left . . . old river": Ash, *Buried Valley*, 11; "when Knox . . . 19 inches": Wolensky, Wolensky, and Wolensky, *Knox*, 71; "At the point . . . buried valley": Wolensky, Wolensky, and Wolensky, *Knox*, 74; "District 1 . . . the business": Wolensky and Hastie, *Anthracite Labor Wars*, 179; "offering evidence . . . close cooperation": Wolensky and Hastie, *Anthracite Labor Wars*, 385, 386; "Courts . . . violations": Wolensky and Hastie, *Anthracite Labor Wars*, 180, 304n43. The miners' supervisors ignored maps that warned of the danger (Miller and Sharpless, *Kingdom of Coal*, 321).

17. "Dewatering . . . coal business": Wolensky, Wolensky, and Wolensky, *Knox*, 110; "With local . . . state average": Wolensky, Wolensky, and Wolensky, *Knox*, 69; "the regional . . . payroll": Miller and Sharpless, *Kingdom of Coal*, 321; "Compounding . . . subsidence": Wolensky, Wolensky, and Wolensky, *Knox*, 69–70.

18. "In 1915 . . . angrily": "Three Hundred Perish"; "drowning . . . colliery": E. Roberts, "A History of Land Subsidence," 5; "After . . . ocean tides": "Drowning of Higashimisome Colliery"; "In 1995 . . . 74 workers": Michalski, "The Jharia Mine Fire," 85n2; "Eroding . . . the fires": Michalski, "The Jharia Mine Fire," 85.

19. "Petroleum companies . . . deny climate change": Weissert, "Big Oil Seeking Protection"; "funnel billions . . . to coal companies": Maslin, *Climate Change*, 155. Even prior to Harvey, Houston had been sinking one to two inches each year (LeMenager, *Living Oil*, 131). Observers noted in 1978 that areas around the Houston Ship Channel had dropped 10 feet due to groundwater withdrawals (Melosi, *Sanitary City*, 232).

20. "the Lackawanna River . . . both banks": "Great Freshet on the Lackawanna"; "Rainfall . . . workers": Durfee, *Reminiscences of Carbondale, Dundaff, and Providence*, 32–34; "another badly . . . as eight": Durfee, *Reminiscences of Carbondale, Dundaff, and Providence*, 71.

21. "The Lack-of-water": "After the Big Flood"; "washeries . . . bottom lines": Healey, *Pennsylvania Anthracite Coal Industry, 1860–1902*, 217.

22. "the first two . . . railroad tracks": "Valley Is Swept by a Terrific Storm," "People Flee from Homes," and "Another Disastrous Flood," 5; "In the . . . tipped over": "Another Disastrous Flood," 12; "men dynamited . . . dams": "Reservoirs at Scranton Gas Works Saved by Dynamite"; "water overwhelmed . . . with runoff": "Great Damage in the Path of the Flood," 6, and "Another Disastrous Flood," 12; "Observers . . . generations": "Editorial Notes," *Scranton Times*, 3 Mar. 1902.

23. "The 1903 . . . 48 hours": "Raging Flood Leaves Wreck," 1; "Reaching . . . 250 families": "Scranton's Greatest Flood," 5, and "Water Recedes"; "As men . . . pumps": "Scranton's Greatest Flood," 9; "Touching . . . Mayfield": "Flood Stops Traffic," 7; "in Olyphant . . . channel": "Raging Flood Leaves Wreck," 3; "and in Carbondale . . . records": "Scranton's Greatest Flood," 5; "the Lackawanna . . . coal cars": "Raging Flood Leaves Wreck," 1; "After . . . anthracite": "Gathering Winter Coal Supply."

24. "Many blamed . . . banks": "Encroaching on the River Banks"; "After the 1902 . . . practice": "First Official Step," 5; "among them . . . Company": "Widen River Channel," 11; "By 1903 . . . 50 feet wide": "Encroachment upon River Channel," 7; "Although . . . the river": "Flushing of Culm"; "people . . . banks": "Where Responsibility for Flood Damages Lies." See "Flushing of Culm" for a photo of the river, the Sauquoit mill, a culm bank, and railroad tracks.

25. "In 1892 . . . supreme court": "That Slag in the River" and "The Supreme Court"; "In 1899 . . . in 1857": "River Bed Has Changed"; "After . . . riverbed": "First Official Step," 3; "ordinances . . . offenders": "Where Responsibility for Flood Damages Lies."

26. "Flushing of Culm."

27. As thoughtlessly, in 2018 the world cast into the atmosphere "about 1,300 tons of carbon dioxide every second" (Borenstein, "Climate Reality Check"). In both cases, polluters expected waste to disappear; instead, it piled up. "In 1902 . . . 20 percent": Leighton, *Quality of Water*, 30; "In 1903 . . . overflow": "Where Responsibility for Flood Damages Lies"; "In 1944 . . . Lackawanna": Ash, Dierks, and Miller, *Mine Flood Prevention*, 74; "Collieries . . . per second": Leighton, *Quality of Water*, 30; "altering . . . to others": Brubaker, *Down the Susquehanna to the Chesapeake*, 58.

28. "Scranton's . . . upon it": "Future of the City"; "D & H . . . culm": "Where Responsibility for Flood Damages Lies"; "to build . . . flooding": "A Retaining Wall."

29. "In mid-March . . . out of work": "Flood Control Omitted Mines," 1; "Finding . . . its banks": "Flood Closes Schools," 1; "the river . . . the current": "Crater Gouged by Flood Waters," see caption; "The high . . . inundating mines": "Flood Control Omitted Mines," 1; "Immediately . . . dry": "Army Experts Study Floods," 3; "The flood . . . tools": "Tools Lost in Mining Flood Worth $100,000"; "In places . . . 1902 crest": "How Army Engineers' Map"; "Contributing to . . . the river": "Seek $305,000,000 for Flood Prevention" and Brubaker, *Down the Susquehanna to the Chesapeake*, 64.

30. "Other floods . . . 42 feet": National Weather Service, National Oceanic and Atmospheric Administration, "Historical Floods," and see Hasco, "These Are the 10 Biggest Floods the Susquehanna River Has Ever Seen."

31. "The average . . . 38 inches": National Weather Service, National Oceanic and Atmospheric Administration, "Normals for Scranton/Wilkes-Barre"; "In 2018 . . . 61 inches": National Weather Service, National Oceanic and Atmospheric Administration, "Local Month/Year . . . 2018"; "The next . . . above average": National Weather Service, National Oceanic and Atmospheric Administration, "Local Month/Year . . . 2019."

32. "For every . . . 7 percent": Romm, *Climate Change*, 48; "Said another . . . more often": Borenstein, "Warming Means Wetter"; "in the Northeast . . . increasing precipitation": Romm, *Climate Change*, 49–51.

33. "Disturbing . . . 30 feet": "Black Thursday," 4; "Near Stroudsburg . . . near 80": Flannery, "Flood Death Toll May Hit 79," 1; "At 88 . . . Pocono Mountains": "Black Thursday," 4; "In Scranton . . . tunnels": "Black Thursday," 13; "not far . . . never rebuilt": Conover, *Labeling Lackawanna*, 58, 113; "Diane . . . to merge": Gallagher, "Hurricane Diane"; "Lessons . . . bankruptcy": "Erie Lackawanna Files for Bankruptcy." For photos of Diane damage, see "Disaster Strikes" and "Then & Now," *Scranton Sunday Times*, 5 Mar. 2017.

34. Inners, LaRegina, and Chamberlin, "'King Coal' on the Mountain."

35. "In the late . . . repeated": F. Adams, "Hurricane Agnes," 30; "an attitude . . . 1902 rise": "Editorial Notes," *Scranton Times*, 3 Mar. 1902; "Too many . . . Susquehanna": Kneeland, *Playing Politics with Natural Disaster*, 123; "for several . . . stretch": Bailey, Patterson, and Paulhus, *Hurricane Agnes Rainfall and Floods, June–July 1972*, 45, 54; "Covering . . . city": Bailey, Patterson, and Paulhus, *Hurricane Agnes Rainfall and Floods, June–July 1972*, 83; "unburied . . . cemetery": Commonwealth of Pennsylvania, Department of Military and Veteran Affairs, Office of the Adjutant General, *After Action Report*, D-11; "scattering . . . on roofs": Kneeland, *Playing Politics with Natural Disaster*, 37; "At the time . . . 117": Bailey, Patterson, and Paulhus, *Hurricane Agnes Rainfall and Floods, June–July 1972*, 1, 83.

36. "In 1973 . . . major rivers": F. Adams, "Hurricane Agnes," 33–34; "Remembering . . . with water": Kneeland, *Playing Politics with Natural Disaster*, 131.

37. "get among . . . stars do": Frost, "The Prerequisites," 418.

38. References to the poem "Stay Home" are cited in text by line number. See Berry, "Stay Home." For "The Pasture," see Frost, "The Pasture."

39. "In 2002 . . . mine waste": Hopton and Vernon, "From the Sublime to Destruction."

40. "Prior to . . . Kentucky": B. Johnson, *Carbon Nation*, 165; "Slurry . . . streams": Leung, "A Toxic Cover-Up?" and Hopton and Vernon, "From the Sublime to Destruction"; "In Tennessee . . . sludge": Biello, "Toxic Ash Pond Collapses in Tennessee"; "In 2019 . . . Atlantic Ocean": Darlington, "Brazil Arrests 8 at Mining Giant Vale," and Kaiser and Silva De Sousa, "Dam Collapse Proves Deadly."

CHAPTER 7

1. "Dams . . . globe": Pearce, *When the Rivers Run Dry*, 133; "After . . . escalated": McNeill and Engelke, *Great Acceleration*, 91; "uprooting . . . people": Pearce, *When the Rivers Run Dry*, 135; "with . . . China": Webber, *Thirst for Power*, 66; "Behind this . . . and more": Schmidt, *Water*, 110; "Although big . . . promise": Smil, *Energy*, 195. Over 50,000 big dams stand around the world (Schmidt, *Water*, 110). The United States has "more than 75,000 dams and reservoirs . . . two million miles of pipe . . . [and] more than 160,000 drinking water facilities" (Salzman, *Drinking Water*, 162).

2. "The world's . . . 200,000": Pearce, *When the Rivers Run Dry*, 149; "In 2020 . . . evacuate": "Dire Warning"; "Two . . . thousands": Magdy, "A Disaster"; "Heavy . . . 2017": "Dire Warning."

3. "industrialization . . . commodity": Steinberg, *Nature Incorporated*, 16, 85; "The poor may go thirsty": Hauter, *Frackopoly*, 162.

4. "Economists . . . wisely": Wharton School of the University of Pennsylvania, "Ebb without Flow."

5. "In 1992 . . . economic good": *The Dublin Statement and Report of the Conference*, 4; "Viewed . . . approaches": *The Dublin Statement and Report of the Conference*, 15; "In 2008 . . . aquifer": Wharton School of the University of Pennsylvania, "Ebb without Flow"; "In 2012 . . . Conoco-Phillips": Hauter, *Frackopoly*, 161; "to figure . . . scarcity": Hauter, *Frackopoly*, 163–164.

6. "In 2000 . . . change": Wharton School of the University of Pennsylvania, "Ebb without Flow"; "declaring . . . commercial purposes": Salzman, *Drinking Water*, 211; "In Michigan . . . to us": Salzman, *Drinking Water*, 19; "In 2002 . . . human right": Salzman, *Drinking Water*, 222; "It should . . . right": Hauter, *Frackopoly*, 165; "In a sign . . . privatize water": Pope Francis, *On Care for Our Common Home*, 20; "in 2017 . . . Sanitation": Pontifical Academy of Sciences, "Final Statement of the Workshop on The Human Right to Water."

7. "With the . . . drinking water": Salzman, *Drinking Water*, 22; "In a few years . . . water stress": Hauter, *Frackopoly*, 164.

8. "dry times . . . Central America": Maslin, *Climate Change*, 112; "In 2006–2010 . . . farmers": Romm, *Climate Change*, 129; "In 2018 . . . go dry": Onishi and Sengupta, "In South Africa, Facing 'Day Zero' with No Water"; "In the Czech Republic . . . cry": "Drought Reveals 'Hunger Stones.'"

9. "global . . . assumed": Webber, *Thirst for Power*, 51; "As . . . 2011–2019": National Integrated Drought Information System, "California"; "irrigated . . . withdrawals": Lund et al., "Lessons from California's 2012–2016 Drought," 3; "In 2023 . . . approved": Rojanasakul et al., "America Is Using Up"; "Beneath . . . production": Gleick, "Water and Energy," 69; "water mining . . . reserve": Postel, "Water and Agriculture," 59.

10. "Places . . . evaporation": Romm, *Climate Change*, 100; "By 2035 . . . same mark": Kummer, "Here's What Climate Change Will Look Like in Northeast."

11. "Defined then . . . 100 years": Water Supply Commission of Pennsylvania, *Report of the Water Supply Commission of Pennsylvania, 1908*, 52; "a dry spell . . . April": "A Crisis Almost Here."

12. "In Indiana . . . bushels aboard": "Water Famine in the West"; "By end of October . . . gone dry": "Railroad and Mine"; "The railroad . . . 5,000": "Carrying Water from Hollidaysburg"; "As . . . water trains": "A Crisis Almost Here."

13. "A Crisis Almost Here."

14. "Mine operators . . . purposes": "Money, Mine and Labor"; "turned to using 'sulphur water'": "Stockton Colliery Men Idle"; "Acidic . . . wear out": "A Big Water Scheme"; "As even . . . closed": "Colliery Men Idle"; "although . . . water companies": "A Big Water Scheme"; "In Carbondale . . . gravity railroad": "Carbondale: Work of the Drought"; "Low water . . . canal": "Honesdale"; "By mid-September . . . elsewhere": "Water Famine"; "In Wilkes-Barre . . . customers": "Plenty of Water"; "By late September . . . lost work": "The Water Famine"; "In early . . . days more": "A Two Days' Rain Needed."

15. "D & H . . . third": "Selling on Old Orders"; "prices . . . a ton": "Coal and Mines"; "With coal . . . want of water": "Coal Prices Advanced."

16. "Although . . . records": "Broke All Records"; "In Scranton . . . reservoirs": "Scranton's Water Supply"; "A flood . . . South Side": "Large Reservoir Burst" and "D., L., W. Yards Flooded"; "November rains . . . November 4": "Pennsylvania Collieries Resume."

17. "To cope . . . Pottsville": "New Dam Built"; "Shamokin . . . reservoirs": "Shamokin's New Water Supply"; "Backers . . . the idea": "A Big Water Scheme."

18. The novel increased the capacity of some readers to imagine climate change and its potential effects (Schneider-Mayerson 344). References to the novel *The Water Knife* are cited in-text by page number. See Bacigalupi, *The Water Knife*.

19. "After crews . . . to settle": Lowenthal, *From the Coalfields*, 64; "From its start . . . sawmill": Lowenthal, *From the Coalfields*, 66; "When Torrey . . . future time": Lowenthal, *From the Coalfields*, 69; "Guarding . . . repair their dam": Lowenthal, *From the Coalfields*, 58; "the D & H . . . moles": LeRoy, *Delaware and Hudson Canal*, 83; "in the red until 1840": Mathews, *History of Wayne, Pike and Monroe Counties*, 243.

20. "110 total": Lowenthal, *From the Coalfields*, 71; "releasing . . . in New York": Lowenthal, *From the Coalfields*, 146; "In 1852 . . . November": Delaware and Hudson Canal Company, *Annual Report . . . 1852*, 4; "The next year . . . under control": Delaware and Hudson Canal Company, *Annual Report . . . 1853*, 3; "In 1854 . . . boats": Lowenthal, *From the Coalfields*, 304.

21. "By 1880 . . . reservoirs": LeRoy, *Delaware and Hudson Canal*, 79; "In Wayne County . . . Swamp": Mathews, *History of Wayne, Pike and Monroe Counties*, 586.

22. Scholars have recently coined several words that have helped me to see more clearly, among them "Anthropocene" (Crutzen and Stoermer, "The Anthropocene"), "hyperobject" (Morton, *Hyperobjects*), and "slow violence" (Nixon, *Slow Violence*).

23. "opened in 1870": Lowenthal, *From the Coalfields*, 234; "lay its line . . . east branch": Trails Conservation Corporation, *Upper Lackawanna Watershed Management Plan*, map, 131.

24. "Chartered in 1887": "City Notes"; "Valley Water . . . Stillwater": "Another Big Water War"; "The company projected . . . miles long": "Common Council"; "When Valley Water . . . 1889": "Proposed Water Works"; "taxpayers . . . price": "Another Big Water War."

25. "the D & H . . . Stillwater": "Carbondale"; "The D & H objected . . . thousands of dollars": "Another Big Water War"; "almost . . . the site": "Source of the Lackawanna"; "When . . . end work": "Another Big Water War"; "Noting . . . take land": "Source of the Lackawanna"; "In response . . . 1896": "Time Extension Refused."

26. "In 1889 . . . their water": "Scranton Wants the Earth"; "Others said . . . supplied Scranton": "The Water Not Good"; "While Carbondale officials . . . thing of beauty": "Carbondale News."

27. "With directors . . . farmland": "The Dam at Stillwater"; "drew water . . . submerged ground": "Uniondale"; "By 1899 . . . Company": "Officers Were Elected." Lackawanna Valley Water Supply Company was chartered on October 1, 1897, with offices in Scranton; its directors included C. R. Manville, Carbondale; A. H. McClintock, Wilkes-Barre; and C. S. Weston, Scranton ("Capitol Hill" and "Rates Are To [*sic*] High," 10).

28. "In the 1930s . . . Rondout Creek": Melosi, *Sanitary City*, 138; "In the 1990s . . . streams and rivers": Salzman, *Drinking Water*, 266; "In 2020 . . . Lehigh River": "Dam Study Timely, Wise." The D & H canal met the Hudson near its confluence with Rondout Creek.

29. "In 1895 . . . Gas and Water": "Four Companies after It." Meanwhile, the city of Scranton considered creating a municipal waterworks ("Four Companies after It").

30. "With . . . Lehigh watershed": "Increased Water Supply"; "When SGW . . . Citizens' Water": "Fight over the Lehigh"; "the latter . . . guard": "Four Companies after It"; "vowed . . . reservoir site": "Water Company's Scheme"; "the SGW plan . . . businesses": "Lehigh River Water"; "the Wilkes-Barre water companies": "Fight over the Lehigh"; "SGW claimed . . . severe drought": "Lehigh River Water."

31. "Lehigh Coal . . . canal": "The Lehigh River Dams"; "SGW argued . . . in the Lehigh watershed": "The Water Supply of Scranton"; "In 1896 . . . with SGW": "Gas and Water Co. Wins"; "pumped . . . in Scranton": T. Murphy, *Jubilee History*, vol. 1, 128.

32. "In early 1896 . . . along its banks": "That Gigantic Scheme"; "a year later . . . Wyoming Valley": "No More Water Famines." To further guard against drought, the company also planned three reservoirs ("No More Water Famines").

33. "To sustain . . . Owens River": Reisner, *Cadillac Desert*, 86; "sucking dry . . . farms": Reisner, *Cadillac Desert*, 62; "in 1923 . . . National Park": "Hetch-Hetchy Dam Dedicated." Melosi discusses the drawing of water from the Owens Valley to Los Angeles (*Sanitary City*, 88–89 and 138–139). A drought in Southern California in 1892–1904 spurred the initial move (Melosi, *Sanitary City*, 88).

34. "These dams . . . 1970s": McNeill and Engelke, *Great Acceleration*, 33–34; "More than . . . worldwide": Pearce, *When the Rivers Run Dry*, 133; "became global . . . international attention": McNeill and Engelke, *Great Acceleration*, 33–34.

35. "By 1941 . . . reservoir": "Senate Added 12 Millions [*sic*] to Flood Bill"; "No money . . . for days": "Honesdale and Hawley Hard Hit," 20; "When . . . in 1947": "Heavy Damage"; "people . . . price": "Decision to Abandon" and "Army Engineers Veto." The Lackawaxen watershed suffered far worse: at least 23 people died in Honesdale ("Honesdale and Hawley Hard Hit," 3). Flooding hit much of eastern Pennsylvania. Doyle & Roth was then American Welding.

36. "Soon . . . too much": Trails Conservation Corporation, *Upper Lackawanna Watershed Management Plan*, 55–56; "In 1957 . . . Union Dale": "Looking Back," 2016; "the next . . . began": "Dynamite Blast Begins Work."

37. "Tennessee . . . 1955": Rogers, "Wayne Will Get Economic Transfusion."

CHAPTER 8

1. "Prior to . . . or culm": P. Roberts, *Anthracite Coal Industry*, 224; "banks of . . . output": Lee, "The Recovery of Anthracite from Culm Banks," 720; "In the early . . . steam": P. Roberts, *Anthracite Coal Industry*, 223; "By 1889 . . . manufacturers": Hitchcock, *History of Scranton*, 430; "new marketable . . . usable": P. Roberts, *Anthracite Coal Industry*, 224–225.

2. "Making their . . . it recovered": Lee, "The Recovery of Anthracite from Culm Banks," 720–721; "By 1899 . . . field": P. Roberts, *Anthracite Coal Industry*, 223; "widening . . . deep mining": Healey, *Pennsylvania Anthracite Coal Industry, 1860–1902*, 217;

"Although . . . to run": Lee, "The Recovery of Anthracite from Culm Banks," 721; "owners . . . sizes": Healey, *Pennsylvania Anthracite Coal Industry, 1860–1902*, 217.

3. "At waste . . . in another": P. Roberts, *Anthracite Coal Industry*, 225, and "Working Anthracite Culm Piles," 569; "Slate . . . or slush": Griffen, "Slush Problem in Anthracite Preparation," 514; "Some companies . . . boreholes": Lee, "The Recovery of Anthracite from Culm Banks," 721, and P. Roberts, *Anthracite Coal Industry*, 221; "Collieries . . . waste": Taber and Taber, *The Delaware, Lackawanna, & Western Railroad*, vol. 1, 252; "some washed . . . beds": Leighton, *Quality of Water*, 26–27.

4. "Workers . . . fresh-mined anthracite": P. Roberts, *Anthracite Coal Industry*, 223; "In 1892 . . . 2,567,335 tons": Anthracite Coal Strike Commission, *Report to the President*, 25; "a 27-fold . . . in 1906": Lee, "The Recovery of Anthracite from Culm Banks," 722; "In 1900 . . . per day": P. Roberts, *Anthracite Coal Industry*, 226; "mine laborers . . . $2.50": P. Roberts, *Anthracite Coal Industry*, 111.

5. "they were . . . carloads": "The Coal Washeries"; "In parades . . . the miners": P. Roberts, *Anthracite Coal Industry*, 206; "Although labor . . . from there": "Olyphant"; "In late . . . wash coal": "Tried to Wreck a Train"; "On October 22 . . . Empire Washery": "Two Serious Riots"; "2,000 . . . widows": "Very Exciting Day."

6. "When anti-strikers . . . to work": "Washeries to be Closed"; "Some mine owners . . . protect them": "Today the Big Battle Begins"; "Although Pennsylvania Coal . . . pumping purposes": "All Idle at Pittston"; "the D. L. W . . . running": "Switchmen to Act"; "Union officials . . . mine pumps": "All Idle at Pittston"; "closing . . . companies": "Today the Big Battle Begins."

7. "In 1890 . . . systems": Melosi, *Sanitary City*, 83.

8. "Brothers . . . 1840s": B. Folsom, *Urban Capitalists*, 30; "founded . . . 1853": B. Folsom, *Urban Capitalists*, 38; "established . . . 1854": B. Folsom, *Urban Capitalists*, 46n2; "Their cousin . . . until 1872": T. Murphy, *Jubilee History*, vol. 1, 121, and B. Folsom, *Urban Capitalists*, 121; "His son . . . as president": T. Murphy, *Jubilee History*, vol. 1, 119; "in 1881 . . . in 1891": B. Folsom, *Urban Capitalists*, 121; "After . . . SGW": T. Murphy, *Jubilee History*, vol. 1, 121; "expanding . . . 1905": "Consolidated Water Company Sold Out"; "His son Worthington . . . New York": B. Folsom, *Urban Capitalists*, 121.

9. "Death of W. W. Scranton Takes Foremost Citizen from Ranks of Workers."

10. T. Murphy, *Jubilee History*, vol. 1, 387.

11. "The national . . . Remingtons": Hitchcock, *History of Scranton*, 501; "At the front marched W. W.": "A Bloody Business"; "the corps . . . or more": Hitchcock, *History of Scranton*, 502, and T. Murphy, *Jubilee History*, vol. 1, 390; "The next day . . . the city": Hitchcock, *History of Scranton*, 503.

12. "Unlimited in Quantity; Unexcelled in Quality."

13. "With . . . in 1899": "Lake Elmhurst or Dam-Site"; "the company . . . Bridge Reservoir": Conover, *Labeling Lackawanna*, 173; "In 1894 . . . rainless days": "Met the Milk Dealers"; "Completed in 1898": "City and County"; "by 1902 . . . 10 reservoirs": Leighton, *Quality of Water*, 39; "which contribute . . . today": Hollowell and Koester, *Ground-Water Resources of Lackawanna County*, 9.

14. "Surface sites . . . man pumps": "They Won't Be Reinstated" and "Operators Feel That They Have the Advantage"; "Early in . . . by electricity": "To Operate Mines"; "Despite . . . abandoned": "History of Strike"; "rain fell . . . region": "Six Inches of Rainfall."

Unfamiliar tasks led to accidents, however: working with office employees in Avoca on May 20, civil engineer and son of D & H vice president James Dickson died while operating two engines that conveyed culm from a waste bank to a washery ("Awful Death of James R. Dickson").

15. "Although . . . June 2": Anthracite Coal Strike Commission, *Report to the President*, 36, and Hudson Coal Company, *The Story of Anthracite,* 245; "The threat . . . table": "They Won't Be Reinstated"; "To stop . . . ran them": "General Gobin in the City En Route to Olyphant," 7; "The union . . . coal production": "Seven Under Arrest"; "Butler . . . records": "Pittston," *Scranton Republican*, and "Mitchell to be Here."

16. "Depending . . . stopped working": "Both Sides Pleased"; "Strikers had . . . without strikers": "Mitchell's Orders Generally Obeyed"; "From then . . . hostilities": "Today Will Decide"; "After . . . idle": "Mitchell's Orders Generally Obeyed."

17. "On May 24 . . . the job": "Will Not Strike"; "At Erie . . . at work": "Critical Hour Has Arrived"; "On June 24 . . . victory": "Break in Ranks of the Miners." Run by W. W.'s brother, Joseph A., the *Scranton Republican* surmised that engineers who remained at work recalled "the experiences of the big coal strike in 1877" ("Both Sides Pleased").

18. "For the first months . . . bobbin works": "Washery Stopped"; "When the Sauquoit . . . idled": "Sauquoit Mills to Shut Down."

19. "Attention . . . close them": "The Conference Ends"; "Although . . . run pumps": "To See Mr. Mitchell."

20. "Stoked . . . 135,000": Hitchcock, *History of Scranton*, 96; "number . . . 27": Hitchcock, *History of Scranton*, 74; "or about . . . mile": Hitchcock, *History of Scranton*, 100; "In 1904 . . . interests": Perry, *"A Fine Substantial Piece of Masonry,"* 24. In 1900, the Lackawanna and Wyoming Valleys boasted 175 breakers between Carbondale and Nanticoke (Taber and Taber, *The Delaware, Lackawanna, & Western Railroad*, vol. 1, 252). In 1914, Scranton was 19.6 square miles (Hitchcock, *History of Scranton*, 100).

21. "Facing . . . age 22": "John Jermyn's Busy Life at an End"; "landing . . . Scrantons": B. Folsom, *Urban Capitalists*, 173; "Jermyn rose . . . olden days": "John Jermyn's Busy Life at an End."

22. "Jermyn . . . railroad": "John Jermyn's Busy Life at an End"; "one of his . . . water well": Throop, *Half Century in Scranton*, 305; "he was . . . Supply": "Source of the Lackawanna"; "After he died . . . Jermyn & Company": "Will Made by John Jermyn"; "adding . . . Company": "No Water-Rate War Says Mr. Jermyn."

23. "In 1897 . . . Water": "Carbondale Equity Case"; "state chartered . . . customers": "Legal"; "Headquartered . . . Carbondale": "Officers Were Elected"; "Dominating . . . Valley": "Carbondale Equity Case" and "Officers Were Elected"; "Consolidated . . . Ararat": "Officers Were Elected."

24. "In May . . . fees": "Carbondale: May Retaliate"; "by August . . . rival company": "Carbondale: Looks Bad for Consolidated," "Angry in Carbondale," and "Carbondale: Water Talk."

25. "Public water . . . Jermyn Water Company": "After Water Scheme"; "In 1902 . . . bonds": "Against City Water Plant," 8; "a year later . . . ruling": "City Water Project Killed."

26. "Several . . . Water Company": "Carbondale: Latest Water Move"; "E. E. Hendrick . . . home": "$10,000 a Year" and "Carbondale: Looks Bad for Consolidated"; "Klots . . . flow": "Carbondale: Looks Bad for Consolidated"; "Not far . . . per minute":

"Carbondale: 400 Gallons a Minute"; "In 1901 . . . estate": Conover, *Labeling Lackawanna*, 26.

27. "On May 27 . . . the lot": "Mr. Jos. Jermyn Acts Promptly"; "On the Thursday . . . would walk": "Strike Situation Is Reviewed"; "but owners . . . on the job": "Will Not Leave Work"; "At the Saturday . . . pallbearer": "Funeral of John Jermyn."

28. "The first strike . . . in Pittston": "A Riot at Pittston"; "The next day . . . inside": "Revolvers Fired at Pittston"; "owners . . . fill mines": "A Riot at Pittston"; "shots rang . . . Grassy Island washery": "Outbreaks at Four Places."

29. "To stop . . . storage reservoirs": "Dynamiting at the National"; "On August 12 . . . summoned": "Rioting at Throop"; "Responding . . . full blast": Editorial.

30. "When . . . arson": "Bellevue Washery Destroyed by Fire"; "Nonunion . . . tons of coal": "Briggs' Washery Destroyed by Fire"; "The next day . . . the place": "Editorial Notes," *Scranton Times*, 9 Aug. 1902; "Not ten . . . reopen it": "Baer Makes Statement on Strike Situation" and "Washery Consumed by a Mysterious Fire"; "About three . . . burned": "Fire Destroys Small Washery."

31. "Owners . . . camps": "Is Violence Near?"; "Early . . . onsite": "Many Washeries Being Operated by Imported Men"; "The Butler . . . New York": "Pittston City News"; "Erie Coal . . . to get away": "Men Tricked Here."

32. "In Duryea . . . more than a day": "Another Effort to Arbitrate Strike"; "As a crowd . . . more strikers": "Four Strikers Shot in Riot at Duryea" and "Duryea the Scene of Wild Tumult"; "Some . . . No. 6 washery": "A Fiendish Deed"; "a washery . . . with water": "Dam Destroyed to Prevent Washery Working"; "That Lehigh Coal . . . regiment": "Troops Now Here"; "Some owners . . . collieries": "Work at Von Storch"; "In mid-July . . . section": "Taylor."

33. "Despite . . . strike began": "Many Washeries Being Operated by Imported Men"; "Adding . . . mid-month": "Green Ridge"; "by end . . . working": "Seventeen Collieries Now in Operation"; "With violence . . . in the valley": "General Gobin in the City En Route to Olyphant," 1.

34. "active . . . judge": Wolensky and Hastie, *Anthracite Labor Wars*, 72, 409. References to the short story "Good Manners and the Water Company" are cited in text by page number. See Johnson, "Good Manners and the Water Company."

35. The narrator also describes the agent as "the water company" (58), not an employee of it. "monopoly . . . water": Wolensky and Hastie, *Anthracite Labor Wars*, 2.

36. "Water Rates Again."

37. "SGW responded . . . into the city": "Water Rates Again"; "in 1899 . . . purchased": "Rates Are To [*sic*] High," 10.

38. "Confirming . . . resistance": "New Water Company for City of Scranton"; "SGW went . . . to supply the city": "Water Co. Gets Charter"; "whether . . . exclusive": "Hearing Was Continued" and see "New Company 'Stands Ready'"; "a claim . . . two water companies": "New Company 'Stands Ready,'" 3; "the state . . . 1903": "City Water Co. Gets Charter."

39. "founder . . . Connell": "Water Co. Gets Charter"; "After . . . 1896": "Then & Now," 2019; "In December . . . city-owned system": "Views on the Water Plant." Connell sold the paper in 1908 ("Then & Now," 2019). After the death of Joseph A., the papers consolidated, appearing as the *Tribune-Republican* until 1936, when it became the *Scranton Tribune* ("Then & Now," 2019).

40. "Complicating . . . headwaters of the Lackawanna": "New Water Company for City of Scranton"; "They also . . . Scranton": "New Water Project Formally Launched."

41. "When City Water . . . W. W.": "Rates Are To [*sic*] High," 10; "Their June . . . watershed": "Scranton Gas & Water Has Acquired Properties of Consolidated Company"; "from . . . miles": "Consolidated Water Company Sold Out."

42. "In 1907 . . . of the people": "Rates Are To [*sic*] High," 3; "In 1902 . . . $8": "Mr. Scranton on the Water Rates" and "New Company 'Stands Ready,'" 3; "When the city . . . industry rates": "Rates Are To [*sic*] High," 3; "To complainers . . . bridges": "Mr. Scranton Talks," 5–6.

43. The name J. Addison Kohlmesser echoes the name J. Pierpont Morgan, whose combination owned local coal mines and was headquartered in New York (Wolensky and Hastie, *Anthracite Labor Wars*, 2).

44. "A few days . . . Lackawanna River": "Washery Was Shut Down"; "As the 1902 . . . overflowed": "Text of Opinion in Archbald Suit"; "In 1913 . . . swift current": Water Supply Commission of Pennsylvania, *Water Resources Inventory Report*, 67; "Piping . . . culm": Water Supply Commission of Pennsylvania, *Water Resources Inventory Report*, 74. When the 1902 strike ended, the National mine did not reopen immediately; the operator, Scranton Coal, had flooded the mine to extinguish a mine fire ("Status at the Mines").

CHAPTER 9

1. "Since 1950 . . . demands": McNeill and Engelke, *Great Acceleration*, 112; "By 2050 . . . acre-feet": Pearce, *When the Rivers Run Dry*, 271; "Accommodating . . . natural world": Webber, *Thirst for Power*, 36; "since 1970 . . . two-thirds": World Wildlife Fund, "Executive Summary," 6.

2. "fossil fuels . . . power plants": U.S. Global Change Research Program, "Key Message #7"; "In 2011 . . . in the U.S.": Abatzoglou et al., "A Primer on Global Climate-Change Science," 30; "In 2013 . . . industries": U.S. Department of Energy, *The Water-Energy Nexus*, 69.

3. "Operators . . . payrolls": "Big Electric Plant"; "By 1910 . . . 169 miles": Aurand, *Coalcracker*, 40; "Running currents . . . anthracite": Hudson Coal Company, *The Story of Anthracite*, 211, 213.

4. "Big Dam on the Wallenpaupack." The "wild scheme" would also supply light, heat, and power to local towns and cities, from Port Jervis to Wilkes-Barre and Scranton, and create boating and fishing opportunities ("Big Dam on the Wallenpaupack").

5. "In 1910 . . . gallons": "Largest Lake in Pennsylvania"; "Designed . . . daily": "Wilsonville Dam"; "project . . . Simpson": "Charter of Wilsonville Dam Issued."

6. "During . . . the city": Lewis, "The Life and Work of Col. L. A. Watres"; "in the 1902 . . . upper valley": Hitchcock, *History of Scranton*, 518; "helping . . . mid-October": "A Few Collieries Resume," "May Go to Shenandoah," and "Non-Union Man Clubbed to Death"; "crews . . . the city": Lewis, "The Life and Work of Col. L. A. Watres." Watres served in the National Guard 1877–1891 and 1898–1904 (T. Murphy, *Jubilee History*, vol. 2, 761).

7. "In the mid-1870s . . . Company": Hitchcock, *History of Scranton*, 66; "serving . . . governor": T. Murphy, *Jubilee History*, vol. 2, 761; "A founder . . . Supply": "Source of the Lackawanna"; "he was . . . drought": "That Gigantic Scheme"; "In 1916 . . . Scran-

ton": "Water Companies—Lessors"; "In 1928 . . . Service Company": "Pennsylvania Gas, Water Co. Observing 100th Anniversary."

8. "Planning . . . operations": "Lackawanna May Build Huge Dam in Pike County"; "D. L. W . . . site": "Lackawanna Negotiating for the Purchase of Big Watershed Near Hawley"; "After . . . in 1923": "Outlines Plans for Power Development"; "Formed . . . the world": Engel and Ruminski, "Pennsylvania Power & Light Company Records"; "Turning . . . Kingdom": PPL Corporation, "About Us"; "and delivers . . . Kentucky": PPL Corporation, "Our Companies." PPL acquired Scranton Electric Company in 1956 (Engel and Ruminski, "Pennsylvania Power & Light Company Records").

9. "PPL . . . enterprise": B. Johnson, *Carbon Nation*, 114; "observers . . . sites": "Power Service Proves Adequate for All Needs Here" and Pinchot, "Giant Power," 561; "third-party . . . agenda": B. Johnson, *Carbon Nation*, 115; "Pinchot's . . . farmer": Pinchot, "Giant Power," 561; "Private . . . the plan": B. Johnson, *Carbon Nation*, 120.

10. The epigraph is from Doyle & Roth Manufacturing, "Engineering." "land . . . 1825": Walster Corporation to Doyle & Roth Manufacturing Company, 154.

11. "heat transfer . . . condensers": Ronquillo, "Understanding Heat Exchangers."

12. "During . . . convenient": "A Gas Range"; "When W. W . . . here": "Death of W. W. Scranton Takes Foremost Citizen from Ranks of Workers," 2.

13. "English . . . lighting": International Correspondence Schools, "Manufacture of Gas," part 1, 2, and Smil, *Energy*, 109; "The gas . . . delivery": "Principal Kinds of Illuminating Gas," 550–551; "Despite . . . water gas": Tarr, "Toxic Legacy," 110; "a mix . . . monoxide": International Correspondence Schools, "Manufacture of Gas," part 2, 1.

14. "To make . . . luminous": International Correspondence Schools, "Manufacture of Gas," part 2, 1; "Before . . . separator": International Correspondence Schools, "Manufacture of Gas," part 1, 25–26, and "Local Gas Plant Furnishing Light to Big Clientele"; "To ensure . . . gas again": "Local Gas Plant Furnishing Light to Big Clientele" and International Correspondence Schools, "Manufacture of Gas," part 1, 27; "By . . . dyes": International Correspondence Schools, "Manufacture of Gas," part 1, 8, and Tarr, "Toxic Legacy," 112.

15. "Concentrated . . . Massachusetts": Tarr, "Toxic Legacy," 110; "manufactured . . . 1950": Tarr, "Toxic Legacy," 108; "Widespread . . . 1912": "Principal Kinds of Illuminating Gas," 550, 551; "in 1927 . . . feet of gas": "Local Gas Plant Furnishing Light to Big Clientele"; "Low start-up . . . water-gas plant": "Principal Kinds of Illuminating Gas," 551.

16. "In the 1920s . . . natural gas": Hauter, *Frackopoly*, 21; "Producing . . . action": Tarr, "Toxic Legacy," 119; "scientists . . . treatment": Tarr, "Toxic Legacy," 122, 123, 126; "By the 1930s . . . service sites": Tarr, "Toxic Legacy," 132–135.

17. "renewable energy . . . renewable electricity": U.S. Department of Energy, *The Water-Energy Nexus*, 42; "In 2019 . . . generation": Milman, "Renewables Surpass Coal."

18. "In support . . . Scranton": Lesnefsky, "Wind Farm Plan Fans Concerns," A1; "A water-energy loop . . . green energy": "Change of Plans for Vandling Hydropower Project Presented"; "Buried . . . railbed": "Change of Plans for Vandling Hydropower Project Presented" and "Richmondale Hydro Plant Plans Are Moving Forward." In 2018, Merchant Hydro changed its name to Renewable Energy Aggregators. The company also has plans for a pumped storage operation at the Old Forge borehole (Renewable Energy Aggregators, "Old Forge"). Construction of the Richmondale project is scheduled to begin in 2024 ("Richmondale Pumped").

19. "Akin . . . the grid": O'Connell, "Hydroelectric Plan Takes a Step," A5; "some can . . . scale": Smil, *Energy*, 199; "With an upper . . . 75 acres": Cassell, "Merchant Hydro Seeks Permit on MW Pumped Storage Project in Pennsylvania"; "By 2050 . . . to five": Romm, *Climate Change*, 244.

20. "A Union Dale . . . real place": Young, "Protect Lackawanna River's Natural Integrity"; "An EPA. . . mine voids": O'Connell, "Hydroelectric Plan Takes a Step," A5.

21. "In 1916 . . . Scranton Grow": "3,500 Lights in New Sign of Electric Co."; "the new lights . . . the U.S.": Pennsylvania Historical and Museum Commission, "Pennsylvania Historical Marker Search"; "A marketing . . . signage": "3,500 Lights in New Sign of Electric Co."

22. "Eleven . . . limelight": Martin, "The Authors Carnival of 1886," 1; "In addition . . . dynamo": "The Authors' Carnival," *Scranton Republican*, 25 Apr. 1886; "The folks . . . elites": "The Authors' Carnival," *Scranton Republican*, 23 Apr. 1886; "Mrs. William . . . *Nights*": Martin, "The Authors Carnival of 1886," 7; "Later . . . Continent": "Henry M. Stanley"; "could . . . network": "The Trial Trip."

23. "Between . . . merge": Engel and Ruminski, "Pennsylvania Power & Light Company Records"; "In 1907 . . . the city": "Scranton to Be Made a City of Lights by Combine"; "By 1910 . . . electricity": "Perseverance Means Success"; "an illusion of clean": B. Johnson, *Carbon Nation*, 107.

24. "In the Marcellus . . . interconnection": Hauter, *Frackopoly*, 169; "Pennsylvania . . . Jessup": O'Connell, "Power Plant Would Create Its Own Power"; "After a court . . . Creek": Wind, "Plant Won't Discharge Wastewater into Creek," A1. Even after accounting for drilling, natural gas–fired power plants use 50 percent less water than coal-fired ones (Webber, *Thirst for Power*, 156) and in 2012 produced, for the first time, more electricity than coal-burning plants (Webber, *Thirst for Power*, 155; Gold, *The Boom*, 248).

25. "suffered . . . in 2017": O'Connell, "Vapor, Subsidence Concern Some Near Power Plant," A11; "where crews . . . in 2018": Hughes, "Reimagine Future," photo caption; "begun in 2004 . . . gallons of water": Lesnefsky, "Millions Pumped into Dolph Mine Fire," A1, A11.

26. "In 2012 . . . County": Fontana, "Transload."

27. "Before it . . . more water": Natural Gas Supply Association, "Processing Natural Gas"; "Compressor . . . for leaks": Natural Gas Supply Association, "The Transportation of Natural Gas"; "a 19-day . . . plant": "Gas Line Tour Included Pipe Installation Work."

28. "Pipelines . . . distances": Webber, *Thirst for Power*, 92, and Finkel, *Pipeline Politics*, 74; "they offer . . . reality": MacLeod, "Holding Water in Times of Hydrophobia," 278; "In 2000 . . . 12": National Transportation Safety Board, "Natural Gas"; "In 2010 . . . 8": "NTSB Releases Report on Explosion"; "In the same year . . . billion": Mitchell, *Carbon Democracy*, 265; "The United States . . . barrels": Webber, *Thirst for Power*, 92; "Between . . . 27 percent": Finkel, *Pipeline Politics*, 62. The U.S. Department of Transportation tracks "significant pipeline incidents"; these cause deaths or injuries, cost more than $50,000, release at least 50 barrels of liquid, and/or cause fires or explosions (Webber, *Thirst for Power*, 92).

29. "Gas Explosion over Mined Area Brings Death and Injuries."

30. "in 2008–2016 . . . plants": Hauter, *Frackopoly*, 176; "in 2011 . . . reservoir": McConnell, "Gas Pipeline," D1. One line among several into New York daily carries "800 million barrels of fracked gas" (Hauter, *Frackopoly*, 184).

CHAPTER 10

1. "To ready . . . vapors": Bulfinch, *Golden Age of Myth and Legend*, 364; "Inspired . . . petitioners": Morford and Lenardon, *Classical Mythology*, 245–246.

2. "The shrine . . . surface": Etiope et al., "The Geological Links of the Ancient Delphic Oracle (Greece)," 821; "The spot . . . emotions": Etiope et al., "The Geological Links of the Ancient Delphic Oracle (Greece)," 824.

3. "Promoters . . . future": Hauter, *Frackopoly*, 198; "Arguing . . . economies": Gold, *The Boom*, 306; "scientists . . . response": Lenton et al., "Climate Tipping Points—Too Risky to Bet Against."

4. "In 2011 . . . drilling": Hauter, *Frackopoly*, 198–199; "Concluding . . . coal": Hauter, *Frackopoly*, 199, and Gold, *The Boom*, 235; "they claimed . . . myth": Wilber, *Under the Surface*, 219; "by 2014 . . . casings": Hauter, *Frackopoly*, 203; "the industry . . . nationwide": Gold, *The Boom*, 308–309. An analysis of company data submitted to the Pennsylvania Department of Environmental Protection in 2014–2018 led Ingraffea and his coresearchers to suspect that reports from oil and gas companies significantly underestimated "total emissions from active wells in the state" (Ingraffea et al., "Reported Methane Emissions from Active Oil and Gas Wells," 5788).

5. "Over a 100-year . . . 86 times more": Romm, *Climate Change*, 81; "fugitive . . . thought": Hmiel et al., "Preindustrial CH4 Indicates Greater Anthropogenic Fossil CH4 Emissions."

6. "Although . . . half did": Gold, *The Boom*, 145; "the discovery . . . or fracking": Gold, *The Boom*, 122; "fracking helped . . . in 2016": Beiser, *The World in a Grain*, 120.

7. "Between . . . rest natural gas": Hauter, *Frackopoly*, 6; "In 2011 . . . 170,000 wells": Wilber, *Under the Surface*, 218; "by 2017 . . . drilled": Pennsylvania Independent Oil & Gas Association, "PA Oil and Gas"; "in 2018 . . . 1973": Koenig, "U.S. 'Likely' Has Taken Over as World's Top Oil Producer."

8. "After the rock . . . giant malls": Gold, *The Boom*, 29; "And one . . . contamination": U.S. Department of Energy, *The Water-Energy Nexus*, 16.

9. "The amount . . . wells": Hauter, *Frackopoly*, 165; "Although . . . 5.5 million": Hauter, *Frackopoly*, 3; "Some . . . million gallons": Tabuchi and Migliozzi, "'Monster Fracks'"; "fracturing . . . gallons": U.S. Department of Energy, *The Water-Energy Nexus*, 15; "about half . . . regions": Hauter, *Frackopoly*, 165.

10. "making . . . output": Gold, *The Boom*, 30; "In the Marcellus . . . surface": Hauter, *Frackopoly*, 162; "with the average . . . fluids": Webber, *Thirst for Power*, 124; "In 2008–2011 . . . inches deep": Hauter, *Frackopoly*, 170; "In 2015 . . . gallons": J. Matthews, "Treated Hydraulic Fracturing Wastewater"; "Although . . . more gas": Webber, *Thirst for Power*, 124.

11. "At first . . . rivers": Wilber, *Under the Surface*, 127; "but much . . . plants": Wilber, *Under the Surface*, 124; "In 2010 . . . other industries": Wilber, *Under the Surface*, 126; "in 2017 . . . treatment plants": J. Matthews, "Treated Hydraulic Fracturing Wastewater"; "industry . . . on site": Wilber, *Under the Surface*, 127–128.

12. "After 2009 . . . had been": "Is Fracking Linked to Oklahoma Earthquakes?" and Hauter, *Frackopoly*, 98–99; "suffering . . . 2016": Borenstein, "Oklahoma Quakes Tied to Fracking Wastewater"; "Although . . . to it": "Is Fracking Linked to Oklahoma Earthquakes?" This is not solely a U.S. phenomenon. In 2018, Dutch officials blamed

gas extraction for a 3.4 magnitude quake that caused widespread property damage in parts of Groningen province (Corder, "Manmade Quakes Force Dutch to Face Future without Gas").

13. C. Matthews, "The Next Big Bet in Fracking: Water."

14. "haul . . . County": Krawczeniuk, "Freight Railroad on the Right Track," C4; "sand mining . . . business": Pearson, *When the Hills Are Gone*, 3. In 2012, Linde Corporation, operator of the Carbondale site, could draw as much as 905,000 gallons of water from the river each day for use at drilling sites (Fontana, "Transload").

15. "Extending . . . ice sheet": Krawczeniuk, "Freight Railroad on the Right Track," C4, and Benson and Wilson, *Frac Sand in the United States*, 23; "Although . . . sandstone": Benson and Wilson, *Frac Sand in the United States*, 8 and 24; "Used to . . . and well": Benson and Wilson, *Frac Sand in the United States*, 1; "St. Peter . . . silica": Benson and Wilson, *Frac Sand in the United States*, 8; "mainly . . . resistant": Benson and Wilson, *Frac Sand in the United States*, 2 and 6; "Selling . . . ton": Benson and Wilson, *Frac Sand in the United States*, 54; "much . . . mines": Benson and Wilson, *Frac Sand in the United States*, 49. Frac sand mining has sparked debate in Wisconsin. At one public meeting, a West Virginian drew a parallel between sand mining and the Hawks Nest Tunnel disaster, calling Gauley Bridge a "town of the living dead" (Pearson, *When the Hills Are Gone*, 79).

16. "After . . . processed": Pearson, *When the Hills Are Gone*, 9; "require . . . of water": Beiser, *The World in a Grain*, 128; "A washing . . . impurities": Pearson, *When the Hills Are Gone*, 15; "with . . . ponds": Beiser, *The World in a Grain*, 129; "After drying . . . 100 cars": Pearson, *When the Hills Are Gone*, 9–10; "The Delaware-Lackawanna . . . frac sand": Krawczeniuk, "Local Railroad Traffic Up Again," C4.

17. "the novel . . . form": Henry, "Extractive Fictions," 414. Jason Molesky similarly asserts that the novel "withhold[s] the catharsis of revelation" ("Gothic Toxicity," 66). References to the novel *Heat and Light* are cited in text by page number. See Haigh, *Heat and Light*.

18. The image of faucets on fire appears in the movie *Gasland*, which Haigh acknowledges as a source (Haigh, *Heat and Light*, 429). The allusion recurs in the novel (87, 220, 252).

19. "Concrete . . . atmosphere": Watts, "Concrete: the Most Destructive Material on Earth"; "Emissions run . . . coke production": Smil, *Energy*, 115–116.

20. See Bass, "The World Below," 151–152.

21. "A 'wild card' . . . groundwater": Wilber, *Under the Surface*, 91; "Not only . . . unplugged": Wilber, *Under the Surface*, 203, and see Gold, *The Boom*, 231–232; "The state . . . mid-1950s": Hauter, *Frackopoly*, 153–154, and Gold, *The Boom*, 231–232.

22. "drilling for oil . . . composition": Conca, "The Fracking Solution Is a Good Cement Job"; "Pressed to . . . drilled": Hauter, *Frackopoly*, 127, McLean, "The Next Financial Crisis Lurks Underground," and Wilber, *Under the Surface*, 180–181; "Coming at . . . the next": Conca, "The Fracking Solution Is a Good Cement Job"; "As an oil . . . it later": Gold, *The Boom*, 285; "Add to this . . . drainage": Gold, *The Boom*, 147.

23. "Between 2009 . . . problems": Hauter, *Frackopoly*, 203; "In 2009 . . . six wells": McGraw, *The End of Country*, 228–230; "In the first nine . . . the state": Wilber, *Under the Surface*, 220; "In 2012 . . . undocumented one": Gold, *The Boom*, 231; "In 2020 . . . since 2011": Levy, "Pennsylvania Orders Gas Well Plugged in Fight over Methane."

24. Rubinkam, "County, Public Left in Dark about Methane Leak." See Rubinkam article inset, "Industry Says Emissions Cut." In 2015, a well casing failed near Los Angeles, releasing 100,000 tons of methane, the largest leak in U.S. history (Finkel, *Pipeline Politics*, 63).

25. "bad cement . . . faucets": Conca, "The Fracking Solution Is a Good Cement Job"; "In 2007 . . . explosion": Wilber, *Under the Surface*, 120; "In 2011 . . . nearby town": Gold, *The Boom*, 286–287; "In the same . . . County": Gold, *The Boom*, 230. In 2021, Cabot Oil & Gas merged with Cimarex Energy to form Coterra Energy (Coterra, "Cabot Oil").

26. "major . . . United States": Gold, *The Boom*, 305; "Drilling . . . rural counties": Hauter, *Frackopoly*, 153; "By 2017 . . . alone": Finkel, *Pipeline Politics*, 118; "including residents of Fort Worth": Wilber, *Under the Surface*, 313n149; "most . . . nation": Wolensky, Wolensky, and Wolensky, *Knox*, 69.

27. "how . . . at all": Gaydos, "Writing the Soul of a Place." Haigh points out, "And yet, the people I'm writing about aren't helpless. Their choices may be limited by their circumstances, but they do have agency and make choices within a certain range of options" (Gaydos, "Writing the Soul of a Place").

28. "Under . . . the cap": Wilber, *Under the Surface*, 89; "Tests . . . uranium": O'Connell, "Driller Cabot Faces Charges," A5; "Like others . . . event": Wilber, *Under the Surface*, 152.

29. "Methane . . . responsibility": Wilber, *Under the Surface*, 92; "Betting . . . play": Wilber, *Under the Surface*, 80; "Houston-based . . . operations": Wilber, *Under the Surface*, 88; "Between 2006 . . . deliveries": Hauter, *Frackopoly*, 140; "leaving . . . bottled water": O'Connell, "Driller Cabot Faces Charges," A5; "In 2012 . . . levels of methane": Phillips, "Federal Public Health Report Highlights Contaminants in Dimock's Water"; "After declaring . . . at the tap": Hauter, *Frackopoly*, 143.

30. "In 2015 . . . Chesapeake": Hauter, *Frackopoly*, 140; "In 2020 . . . Cabot argued": O'Connell, "Driller Cabot Faces Charges," A5; "article 1 . . . and water": Pennsylvania General Assembly, *Constitution of the Commonwealth of Pennsylvania*; "a provision . . . fish kill": Griswold, *Amity and Prosperity*, 240.

31. "the gas industry . . . oversight": Office of the Attorney General, Commonwealth of Pennsylvania, *Report 1 of the Forty-Third Statewide Investigating Grand Jury*, 64–65; "Not very . . . water wells": Gold, *The Boom*, 231; "Employing . . . incident": Wilber, *Under the Surface*, 133; "Oversight . . . on drilling": Wilber, *Under the Surface*, 82.

32. "After the DRBC . . . sued": Maykuth, "Fracking Suit Sent Back Down"; "Seven . . . basin": Rubinkam, "Delaware River Panel Wants Ban on Fracking"; "A few months . . . should provide": Bonomo, "Clean Water Low State Priority"; "The DRBC . . . 2021": Rubinkam, "Fracking Ban Gets Tougher"; "Still . . . project": Walsh, "Fight over Fracking."

33. "In *Pennsylvania Coal* . . . private pond": Casner, "Acid Mine Drainage," 101; "In 2020 . . . cleaner industries": PennFuture and Conservation Voters of Pennsylvania, "We Should Never . . ."

CHAPTER 11

1. "*all whatness is wetness*": Feldman, "Thoughts on Thales," 6; "He also . . . touched it": Brumbaugh, *The Philosophers of Ancient Greece*, 14; "His home . . . harbor": Tufts University, "Miletus (Site)."

2. "Plato . . . knowledge": Plato, *Theaetetus*, 69–70.

3. "In the early . . . in 1904": Brubaker, *Down the Susquehanna to the Chesapeake*, 129; "For 100 . . . at home": Hitchcock, *History of Scranton*, 81.

4. "Using ash . . . anthracite": Wagner, *The Billy Marks*, 13:34, 14:18, and see Corgan, *Dredging Pennsylvania Anthracite*, 14; "Operating April to October": Wagner, *The Billy Marks*, 10:46; "the coal fleet . . . in 1940": Corgan, *Dredging Pennsylvania Anthracite*, 17; "Rivaling . . . 1920": "Pumping a Million Tons of Coal from River-Beds"; "In these years . . . 2 percent": Corgan, *Dredging Pennsylvania Anthracite*, 19; "In the mid-1950s . . . streams": Wagner, "The Hard Coal Navy."

5. "Susquehanna . . . business": Wagner, *The Billy Marks*, 31:53; "In 1914 . . . tons": Water Supply Commission of Pennsylvania, *Water Resources Inventory Report*, 122; "including . . . 29,000 tons": "Taking Coal from River"; "Dredgers . . . Company": Water Supply Commission of Pennsylvania, *Water Resources Inventory Report*, 122, 126, 130; "In Pittston . . . the Susquehanna": "Taking Coal from River"; "In the 1920s . . . District": Selig, "River Coal for Heating School Buildings," 43; "in 1996 . . . load": Wagner, "The Hard Coal Navy"; "One of . . . Light": Brubaker, *Down the Susquehanna to the Chesapeake*, 127, 128.

6. "In 1914 . . . Lehigh River": Water Supply Commission of Pennsylvania, *Water Resources Inventory Report*, 122; "In the same year . . . condenser": Hitchcock, *History of Scranton*, 82–83; "In 1925–1953": Brubaker, *Down the Susquehanna to the Chesapeake*, 130; "Holtwood . . . tons": Jones, *Routes of Power*, 191.

7. Bulfinch, *Golden Age of Myth and Legend*, 19–21.

8. For definitions of "aftermath," see *Merriam-Webster's*.

9. "Once dead . . . Pennsylvania": Lockwood, "The Water Is Fine:"; "along with . . . piles": McGurl, "Lackawanna River Stormwater Agency Needed."

10. "In 2016 . . . climate change": McGurl, "Saving River Requires Citizen Involvement."

11. "It was . . . now": Zaitz, "Recollections of a Slovenian Miner."

12. "I experienced . . . absentee owners": Zaitz, "Recollections of a Slovenian Miner."

13. "In the 1830s . . . Hankins Pond": Singleton, "Efforts to Save Wayne County Dam Underway"; "In 1918 . . . White Oak": "Looking Back," 2018; "For decades . . . dismantling": Singleton, "Wayne County Wins Injunction against Historic Dam Demolition"; "in 2019 . . . gas industry": Singleton, "Historic Dam Gets Reprieve; No Demolition."

14. "Prior to . . . 400 ppm": Smil, *Energy*, 144; "Today: 419": Scripps Institution of Oceanography, University of California, San Diego, "The Keeling Curve"; "The most widely . . . In 1958": Smil, *Energy*, 144; "Charles . . . hemisphere": McNeill and Engelke, *Great Acceleration*, 74.

Works Cited

Abatzoglou, John T., Joseph F. C. DiMento, Pamela Doughman, and Stefano Nespor. "A Primer on Global Climate-Change Science." In *Climate Change: What It Means for Us, Our Children, and Our Grandchildren*, 2nd ed., edited by Joseph F. C. DiMento and Pamela Doughman, 15–52. MIT Press, 2014.

"An Act to Amend the Act Entitled, 'An Act to Incorporate the President, Managers and Company of the Delaware and Hudson Canal Company.'" 20 Apr. 1825. In *Laws of the State of New York Passed at the Forty-Eighth Session of the Legislature, Begun and Held at the City of Albany, the Fourth Day of January, 1825*, 332. E. Croswell, 1825.

"An Act to Improve the River Lackawaxen." 13 Mar. 1823. In *Acts of the General Assembly of the Commonwealth of Pennsylvania*, 74–84. William Greer, 1823.

Adams, Eryn J., Anh T. Nguyen, and Nelson Cowan. "Theories of Working Memory: Differences in Definition, Degree of Modularity, Role of Attention, and Purpose." *Language, Speech, and Hearing Services in Schools* 49 (July 2018): 340–355. https://doi.org/10.1044/2018_LSHSS-17-0114.

Adams, Franklin S. "Hurricane Agnes: Flooding vs. Dams in Pennsylvania." *Bulletin of the Atomic Scientists* 29, no. 4 (Apr. 1973): 30–34.

"After the Big Flood." *Scranton Republican*, 12 Oct. 1903: 5.

"After Water Scheme." *Scranton Republican*, 25 July 1901: 5.

"Against City Water Plant." *Scranton Tribune*, 29 Jan. 1902: 3, 8.

Alaimo, Stacy. *Bodily Natures: Science, Environment, and the Material Self.* Indiana University Press, 2010.

Albrecht, Glenn, Gina-Maree Sartore, Linda Connor, Nick Higginbotham, Sonia Freeman, Brian Kelly, Helen Stain, Anne Tonna, and Georgia Pollard. "Solastalgia: The Distress Caused by Environmental Change." *Australasian Psychiatry* 15 (2007). https://doi.org/10.1080/10398560701701288.

Aldrich, Mark. "An Energy Transition before the Age of Oil: The Decline of Anthracite, 1900–1930." *Pennsylvania History* 85, no. 1 (Winter 2018): 1–31.

Allabaugh, Denise. "Mining a Milestone." *Citizen's Voice* [Wilkes-Barre], 4 Feb. 2018: A20, A21.
"All Hope at an End." *Wilkes-Barre Evening Leader*, 22 Dec. 1885: 2.
"All Idle at Pittston." *Scranton Republican*, 20 Sept. 1900: 1.
"Angry in Carbondale." *Scranton Republican*, 20 Oct. 1899: 5.
"Another Big Water War." *Scranton Republican*, 14 Dec. 1895: 5.
"Another Disastrous Flood." *Scranton Republican*, 1 Mar. 1902: 5, 12.
"Another Effort to Arbitrate Strike." *Scranton Republican*, 13 Aug. 1902: 5.
Anthracite Coal Strike Commission. *Report to the President on the Anthracite Coal Strike of May-October, 1902*. GPO, 1903. https://books.google.com/books?id=sCryAAAAMAAJ&pg=PA37&source=gbs_toc_r&cad=3#v=onepage&q&f=false. Accessed 27 Mar. 2021.
"Anthracite Mining under Buried Valley Dangerous." *Connellsville Daily Courier*, 28 Mar. 1951: 12.
Apollo, Belvidere. D & H Dam Destroyed. *Susquehanna Democrat*, 8 May 1829: 2.
Appalachian Regional Commission, Department of Environmental Protection. *Relationship between Underground Mine Water Pools and Subsidence in the Northeastern Pennsylvania Anthracite Fields*. Prepared by A. W. Martin Associates, 30 Apr. 1975.
"Archbald's Famous Glacial Pot-Hole." *Scranton Tribune*, 15 July 1899: 11.
"Army Engineers Veto River Flood Control Project." *Scranton Times*, 1 Dec. 1945: 3.
"Army Experts Study Floods." *Scranton Republican*, 25 Mar. 1936: 1, 3.
Ash, S. H. *Buried Valley of the Susquehanna River: Anthracite Region of Pennsylvania*. Bulletin 494, U.S. Department of the Interior, Bureau of Mines. GPO, 1950. University of North Texas Digital Library. https://digital.library.unt.edu/ark:/67531/metadc12653/. Accessed 15 Aug. 2019.
Ash, S. H., H. A. Dierks, and P. S. Miller. *Mine Flood Prevention and Control: Anthracite Region of Pennsylvania, Final Report of the Anthracite Flood-Prevention Project Engineers*. GPO, 1957. University of North Texas Digital Library. https://digital.library.unt.edu/ark:/67531/metadc12717/. Accessed 19 Dec. 2019.
Ash, S. H., et al. Romishcer. *Data on Pumping at the Anthracite Mines of Pennsylvania*. U.S. Department of the Interior, Bureau of Mines, 1950. University of North Texas Digital Library. https://digital.library.unt.edu/ark:/67531/ metadc38541/m1/1/?q=bureau%20of%20mines%20bulletin%203776%20water%20pum ping. Accessed 15 Mar. 2021.
Associated Press. "The EPA Finalizes a Water-Protection Rule That Repeals Trump-Era Changes." NPR. 30 Dec. 2022. https://www.npr.org/2022/12/30/1146355861/epa-water-protections-wetlands-rule. Accessed 24 Oct. 2023.
"As to Public Baths." *Carbondale Leader*, 31 Mar. 1898: 4.
Atkins, Peter. *The Laws of Thermodynamics: A Very Short Introduction*. Oxford University Press, 2010.
Audubon Florida. "Kissimmee River Restoration." https://fl.audubon.org/conservation/kissimmee-river-restoration. Accessed 1 July 2020.
Aurand, Harold. *Coalcracker Culture: Work and Values in Pennsylvania Anthracite, 1835–1935*. Susquehanna University Press, 2013. Repr., Rosemont Publishing, 2003.
———. "The Lattimer Massacre: Who Owns History?—An Introduction." *Pennsylvania History* (2002): 5–10. https://journals.psu.edu/phj/article/viewFile/25735/25504. Accessed 7 May 2024.

"The Authors' Carnival." *Scranton Republican*, 23 Apr. 1886: 3.

"The Authors' Carnival." *Scranton Republican*, 25 Apr. 1886: 5.

"Awful Death of James R. Dickson." *Scranton Tribune*, 21 May 1902: 5.

Bacigalupi, Paolo. *The Water Knife*. Vintage, 2015.

"Bad Bathing Place." *Scranton Republican*, 3 Aug. 1898: 5.

"Baer Makes Statement on Strike Situation." *Scranton Times*, 27 Aug. 1902: 1.

Bailey, J. F., J. L. Patterson, and J. L. H. Paulhus. *Hurricane Agnes Rainfall and Floods, June–July 1972*. Geological Survey Professional Paper 924, U.S. Geological Survey and the National Oceanic and Atmospheric Administration. GPO, 1975. https://pubs.usgs.gov/pp/0924/report.pdf. Accessed 5 Aug. 2020.

Barbe, Walter B., and Kurt A. Reed, eds. *History of Wayne County, Pennsylvania, 1798–1998*. Wayne County Historical Society, 1998.

Barnett, Cynthia. *Rain: A Natural and Cultural History*. Broadway, 2015.

Bass, Rick. "The World Below." In *Fracture: Essays, Poems, and Stories on Fracking in America*, edited by Taylor Brorby and Stephanie Brook Trout, 150–162. Ice Cube Press, 2016.

"Begin Task of Removing Water." *Wilkes-Barre Record*, 15 Apr. 1924: 15.

Beiser, Vince. *The World in a Grain: The Story of Sand and How It Transformed Civilization*. Riverhead, 2018.

Belin, Paul B. "Address to the Scranton Chamber of Commerce." 29 June 1927. Scranton Lace Company file, 677 SCR 15 L116. Archives of the Lackawanna Historical Society.

"Bellevue Washery Destroyed by Fire." *Scranton Republican*, 2 Aug. 1902: 5.

Benson, Mary Ellen, and Anna B. Wilson. *Frac Sand in the United States: A Geological and Industrial Overview*. U.S. Geological Survey, 2015. https://pubs.usgs.gov/of/2015/1107/pdf/ofr20151107.pdf. Accessed 10 Oct. 2020.

Berry, Wendell. "Stay Home." In *Collected Poems, 1957–1982*, 199. North Point Press, 1984.

Biello, David. "Toxic Ash Pond Collapses in Tennessee." *Scientific American*, 23 Dec. 2008. https://www.scientificamerican.com/article/toxic-ash-pond-collapses/. Accessed 20 July 2020.

"Big Dam on the Wallenpaupack." *Milford Dispatch*, 4 Sept. 1902: 1.

"Big Electric Plant." *Scranton Republican*, 6 Nov. 1902: 5.

"A Big Sewer Scheme." *Scranton Times*, 28 Oct. 1891: 1.

"A Big Water Scheme." *Wilkes-Barre Record*, 6 Sept. 1895: 8.

Biro, Andrew. "River-Adaptiveness in a Globalized World." In *Thinking with Water*, edited by Cecilia Chen, Janine MacLeod, and Astrida Neimanis, 166–184. McGill-Queen's University Press, 2013.

Blackman, Emily C. *History of Susquehanna County, Pennsylvania*. Claxton, Remsen & Haffelfinger, 1873.

"Black Thursday." *The Lackawanna* 2, no. 6 (Sept. 1955): 4–7, 13.

"Blamed Their Fathers." Advertisement for Richmont Park. N.d. N.p. Author collection.

"A Bloody Business." *Scranton Weekly Republican*, 8 Aug. 1877: 4.

"Blue Coal Is Here!" *Boston Globe*, 14 May 1929: 10.

"Blue Coal, Pink Coal." *Time*, 9 July 1928. http://content.time.com/time/subscriber/article/0,33009,723485,00.html. Accessed 7 May 2024.

"The Board of Health." *Scranton Republican*, 10 Aug 1889: 5.
Bodnar, John. *Anthracite People: Families, Unions and Work, 1900–1940*. Pennsylvania Historical and Museum Commission, 1983.
"Boil Drinking Water; Boil Milk; Keep Down Typhoid Fever Spread." *Scranton Times*, 27 Aug. 1909: 1.
Bonomo, Jacquelyn. "Clean Water Low State Priority." *Scranton Times-Tribune*, 21 July 2017: A12.
Borenstein, Seth. "Arctic in Midst of Unprecedented Transition." *Scranton Times-Tribune*, 13 Dec. 2018: B7.
———. "The Big Chill? Not This Year." *Scranton Times-Tribune*, 9 March 2018: B7.
———. "Climate Reality Check." *Scranton Times-Tribune*, 6 Dec. 2018: B7.
———. "Hey, Who Left Earth's Refrigerator Door Open?" *Scranton Times-Tribune*, 22 Aug. 2019: B7.
———. "'Life or Death': U.N. Report on Global Warming Carries Dire Warning." *Scranton Times-Tribune*, 9 Oct. 2018: A1, A6.
———. "Melting Antarctica: Continent's Ice Sheet Melting 3 Times Faster." *Scranton Times-Tribune*, 14 June 2018: B7.
———. "More Warmth, More Fires." *Citizen's Voice* [Wilkes-Barre], 20 Aug. 2018: B7.
———. "Oklahoma Quakes Tied to Fracking Wastewater." *Scranton Times-Tribune*, 2 Feb. 2018: B7.
———. "2020 Sets Yet Another Temperature Record." *Scranton Times-Tribune*, 15 Jan. 2021: B10.
———. "Warming Means Wetter." *Scranton Times-Tribune*, 30 Aug. 2017: B7.
———. "Waterways Are Turning Green, and Not with Envy." *Scranton Times-Tribune*, 9 Jan. 2021: B7.
Borenstein, Seth, and Frank Jordans. "It's a Summer of Extremes." *Scranton Times-Tribune*, 28 July 2018: B7.
Borenstein, Seth, and Jamey Keaten. "Waste Land." *Scranton Times-Tribune*, 9 Aug. 2019: B7.
"Both Sides Pleased." *Scranton Republican*, 3 June 1902: 5.
Boyle, Louise. "Biden Reverses Trump Rollback and Tightens Limits on Toxic Pollution from Coal Power Plants." *Independent*, 26 July 2021. https://www.independent.co.uk/climate-change/biden-toxic-wastewater-coal-fossil-fuel-epa-b1890899.html. Accessed 24 Oct. 2023.
Brady, Jeff. "Biden Signs Bill to Restore Regulations on Climate-Warming Methane Emissions." NPR. *Morning Edition*, 30 June 2021. https://www.npr.org/2021/04/28/991635101/congress-votes-to-restore-regulations-on-climate-warming-methane-emissions. Accessed 7 May 2024.
"Break in Ranks of the Miners." *Scranton Times*, 24 June 1902: 1.
"Briggs' Washery Destroyed by Fire." *Scranton Times*, 8 Aug. 1902: 1.
"Broke All Records." *Scranton Tribune*, 1 Nov. 1895: 5.
Brown, William H. *History of the First Locomotives in America*. 1874. http://www.public-library.uk/dailyebook/The%20history%20of%20the%20first%20locomotives%20in%20 America%20(1874).pdf. Accessed 10 Mar. 2021.
Brubaker, Jack. *Down the Susquehanna to the Chesapeake*. Penn State University Press, 2002.

Brumbaugh, Robert S. *The Philosophers of Ancient Greece.* SUNY Press, 1981. Repr., Crowell, 1964.
Buell, Lawrence. *The Future of Environmental Criticism: Environmental Crisis and Literary Imagination.* Blackwell, 2005.
———. *Writing for an Endangered World: Literature, Culture, and Environment in the U.S. and Beyond.* Harvard University Press, 2001.
"A Builder Speaks about the Wheels." *Carbondale Leader*, 8 Feb. 1902: 2.
Bulfinch, Thomas. *The Golden Age of Myth and Legend.* Wordsworth Editions, 1993.
"Capitol Hill." *Harrisburg Daily Independent*, 1 Oct. 1897: 2.
"Carbondale." *Scranton Times*, 14 Dec. 1895: 8.
"Carbondale Equity Case." *Scranton Republican*, 13 Sept. 1901: 3.
"Carbondale: 400 Gallons a Minute." *Scranton Tribune*, 2 Sept. 1899: 9.
"Carbondale: Latest Water Move." *Scranton Tribune*, 21 Nov. 1899: 9.
"Carbondale: Looks Bad for Consolidated." *Scranton Tribune*, 12 Oct. 1899: 9.
"Carbondale: May Retaliate." *Scranton Tribune*, 4 May 1899: 9.
"Carbondale News." *Scranton Republican*, 18 Sept. 1896: 6.
"Carbondale: Water Talk." *Scranton Tribune*, 2 Aug. 1899: 7.
"Carbondale: Work of the Drought." *Scranton Tribune*, 8 Oct. 1895: 8.
"Carrying Water from Hollidaysburg." *Carbondale Leader*, 5 Oct. 1895: 1.
Casner, Nicholas. "Acid Mine Drainage and Pittsburgh's Water Quality." In *Devastation and Renewal: An Environmental History of Pittsburgh and Its Region*, edited by Joel A. Tarr, 89–109. University of Pittsburgh Press, 2003.
Cassell, Barry. "Merchant Hydro Seeks Permit on MW Pumped Storage Project in Pennsylvania." Power Grid International. 12 Jan. 2017. https://www.power-grid.com/der-grid-edge/merchant-hydro-seeks-permit-on-230-mw-pumped-storage-project-in-pennsylvania/. Accessed 23 Mar. 2021.
"Cave after Water." *Wilkes-Barre Record*, 30 May 1899: 5.
"Cave in Bed of Brook Diverts Water to Mine." *Scranton Times*, 8 Sept. 1917: 1.
"Chance Cone Cleaned Coal." Advertisement. In *Mac's Directory and Handbook of Anthracite*, 26. Coal Information Bureau, 1937.
Chandler, Alfred D., Jr. "Anthracite Coal and the Beginnings of the Industrial Revolution in the United States." *Business History Review* 46, no. 2 (Summer 1972): 141–181.
"Change of Plans for Vandling Hydropower Project Presented." *Forest City News*, 4 Mar. 2020: 2.
Chapman, Isaac A. "Description of Luzerne County." *Hazard's Register of Pennsylvania* 6, no. 7 (14 Aug. 1830): 97–104.
"Charter of Wilsonville Dam Issued." *Scranton Times*, 5 Oct. 1910: 1.
Chen, Cecilia. "Mapping Waters: Thinking with Watery Places." In *Thinking with Water*, edited by Cecilia Chen, Janine MacLeod, and Astrida Neimanis, 274–298. McGill-Queen's University Press, 2013.
Chittick, James. *Silk Manufacturing and Its Problems.* James Chittick, 1913.
"City and County: Brevities." *Scranton Republican*, 24 Nov. 1898: 3.
"City Firm Gets Contract to Halt Flood Menace." *Scranton Times*, 10 May 1962: 3, 8.
"City Notes." *Scranton Tribune*, 24 Dec. 1895: 5.
"City Water Co. Gets Charter." *Scranton Times*, 10 Sept. 1903: 1.
"City Water Project Killed." *Scranton Republican*, 14 Feb. 1903: 10.

Cline, Elizabeth L. *Overdressed: The Shockingly High Cost of Cheap Fashion*. Penguin, 2013.
"Coal and Mines." *Wilkes-Barre Record*, 19 Sept. 1895: 7.
"Coal Mines Reclaimed." *Lebanon Daily News*, 7 Oct. 1895: 1.
"Coal Prices Advanced." *Scranton Republican*, 12 Oct. 1895: 2.
"The Coal Washeries." *Scranton Republican*, 1 Oct. 1900: 5.
"Colliery Men Idle." *Carbondale Leader*, 17 Sept. 1895: 1.
"Common Council." *Scranton Republican*, 17 Aug. 1889: 5.
Commonwealth of Pennsylvania, Department of Military and Veteran Affairs, Office of the Adjutant General. *After Action Report: The Pennsylvania National Guard, "Tropical Storm Agnes."* Commonwealth of Pennsylvania, Department of Military and Veteran Affairs, Office of the Adjutant General, 1972.
Conca, James. "The Fracking Solution Is a Good Cement Job." *Forbes*, 10 Sept. 2012. https://www.forbes.com/sites/jamesconca/2012/09/10/the-fracking-solution-is-a-good-cement-job/?sh=7e20252e47d1. Accessed 24 Mar. 2021.
———. "Making Climate Change Fashionable—The Garment Industry Takes on Global Warming." *Forbes*, 3 Dec. 2015.
"The Conference Ends." *Scranton Republican*, 23 May 1902: 1.
Conley, Walter R. "Experience with Anthracite-Sand Filters." *Journal (American Water Works Association)* 53, no. 12 (Dec. 1961): 1473–1483.
Conover, Willis. *Labeling Lackawanna: The Story behind Lackawanna County Place-Names*. Lackawanna Historical Society, 2022.
"Consolidated Water Company Sold Out." *Scranton Republican*, 28 June 1905: 3.
Conversion-website.com. "Tons (Water) to Gallons (US Liquid)." http://www.conversion-website.com/volume/ton-water-to-gallon-US-liquid.html. Accessed 12 Jan. 2021.
Corder, Mike. "Manmade Quakes Force Dutch to Face Future without Gas." *Scranton Times-Tribune*, 29 Jan. 2018: A8.
Corgan, Joseph A. *Dredging Pennsylvania Anthracite*. Information circular, U.S. Department of the Interior, Bureau of Mines, June 1942. HathiTrust. https://babel.hathitrust.org/cgi/pt?id=mdp.39015077570862&view=1up&seq=1. Accessed 14 Mar. 2021.
Corlsen, Carl. *Buried Black Treasure: The Story of Pennsylvania Anthracite*. Owen Publishing Company, 1954.
Coterra. "Cabot Oil & Gas and Cimarex Energy Complete Combination, Forming Coterra Energy." 1 Oct. 2021. https://investors.coterra.com/Investors/news/news-details/2021/Cabot-Oil--Gas-and-Cimarex-Energy-Complete-Combination-Forming-Coterra-Energy/default.aspx. Accessed 7 May 2024.
Crable, Ad. "Radioactive Remains." *Scranton Sunday Times*, 31 Mar. 2019: A8.
"Crater Gouged by Flood Waters and Dike Built to Protect Lower Valley Mines." *Scranton Republican*, 25 Mar. 1936: 3.
"A Crisis Almost Here." *Pottsville Miners Journal*, 30 Aug. 1895: 1.
"Critical Hour Has Arrived." *Scranton Tribune*, 2 June 1902: 1.
Crutzen, Paul, and Eugene Stoermer. "The Anthropocene." *IGBP Global Change Newsletter* 41 (May 2000): 17–18.
"Culm in the River." *Wilkes-Barre Times*, 4 Aug. 1900: 5.
"Culm in the River." *Wilkes-Barre Weekly Times*, 11 Aug. 1900: 3.
"D., L., W. Yards Flooded." *Scranton Republican*, 11 Oct. 1895: 3.
Daly, Matthew, and Jill Colvin. "A New Era in Energy Production." *Scranton Times-Tribune*, 29 March 2017: A6.

"The Dam at Stillwater." *Carbondale Leader*, 2 Oct 1897: 6.

"Dam Destroyed to Prevent Washery Working." *Scranton Republican*, 20 Sept. 1902: 9.

"Dam Study Timely, Wise." Editorial. *Scranton Sunday Times*, 12 Jan. 2020: C3.

Dance, Scott. "Chesapeake Bay Foundation to Sue EPA." *Scranton Times-Tribune*, 28 Jan. 2020: B9.

Darlington, Shasta. "Brazil Arrests 8 at Mining Giant Vale over Deadly Dam Collapse." *New York Times*, 15 Feb. 2019. https://www.nytimes.com/2019/02/15/world/africa/brazil-dam-arrests.html?searchResultPosition=2. Accessed 17 Mar. 2021.

Davenport, Coral. "'Climate Change' Mentions Scrubbed from Fed Websites." *Scranton Times-Tribune*, 11 Jan. 2018: B7.

Dearen, Jason, and Mike Schneider. "When Green Isn't Good." *Scranton Times-Tribune* 5 May 2017: B7.

"Death of W. W. Scranton Takes Foremost Citizen from Ranks of Workers." *Scranton Republican*, 4 Dec. 1916: 1, 2.

"Decision to Abandon Stillwater Dam Flood Project Is Reiterated." *Scranton Times*, 7 Aug. 1947: 13.

Delaware and Hudson Canal Company. *Annual Report of the Board of Managers of the Delaware and Hudson Canal Co. to the Stockholders for the Year 1825*. 7 Mar. 1826. https://www.google.com/books/edition/Annual_Report_of_the_Board_of_Managers_o/6CwoAAAAYAAJ. Accessed 20 Mar. 2021.

———. *Annual Report of the Board of Managers of the Delaware and Hudson Canal Co. to the Stockholders for the Year 1837*. 6 Mar. 1838. https://www.google.com/books/edition/Annual_Report_of_the_Board_of_Managers_o/6CwoAAAAYAAJ. Accessed 15 Mar. 2021.

———. *Annual Report of the Board of Managers of the Delaware and Hudson Canal Co. to the Stockholders for the Year 1852*. 29 Mar. 1853. https://www.google.com/books/edition/Annual_Report_of_the_Board_of_Managers_o/6CwoAAAAYAAJ. Accessed 19 Mar. 2021.

———. *Annual Report of the Board of Managers of the Delaware and Hudson Canal Co. to the Stockholders for the Year 1853*. 28 Mar. 1854. https://www.google.com/books/edition/Annual_Report_of_the_Board_of_Managers_o/6CwoAAAAYAAJ. Accessed 19 Mar. 2021.

Dennis, Brady. "'Not a Problem You Can Run Away From': Communities Confront the Threat of Unregulated Chemicals in Their Drinking Water." *Washington Post*, 2 Jan. 2019. https://www.washingtonpost.com/national/health-science/not-a-problem-you-can-run-away-from-communities-confront-the-threat-of-unregulated-chemicals-in-their-drinking-water/2019/01/01/a9be8f72-dd4b-11e8-b732-3c72cbf131f2_story.html. Accessed 20 July 2020.

Dickenson, Wharton. "Brigadier-General Samuel Meredith: First Treasurer of the United States." *Magazine of American History* 3, no. 9 (Sept. 1879): 555–563.

Dierks, H. A., W. L. Eaton, R. H. Whaite, and F. T. Moyer. *Mine Water Control Program, Anthracite Region of Pennsylvania, July 1955–December 1961*. U.S. Department of the Interior, Bureau of Mines, 1962. HathiTrust. https://babel.hathitrust.org/cgi/pt?id=mdp.39015077564246&view=1up&seq=3. Accessed 7 May 2024.

"Dimictic Lake: Vertical Mixing and Overturn." *Encyclopedia Britannica*. https://www.britannica.com/science/dimictic-lake. Accessed 27 Mar. 2021.

"Dire Warning." Editorial. *Scranton Times-Tribune*, 1 June 2020: A10.

"Disaster Strikes, People Rally, Recovery Begins." Flood pictorial supplement. *Scranton Tribune*, 27 Aug. 1955.
Doyle & Roth Manufacturing. "Engineering." http://www.doyleroth.com/engineering.html. Accessed 2 Feb. 2021.
"Drought Reveals 'Hunger Stones.'" *Scranton Times-Tribune*, 25 Aug. 2018: B9.
"Drowing [*sic*] the Schooley." *Wilkes-Barre Leader*, 15 May 1899: 1.
"Drowning of Higashimisome Colliery." *Colliery Engineer* 36, no. 1 (Aug. 1915): 19.
Dublin, Thomas, and Walter Licht. *The Face of Decline: The Pennsylvania Anthracite Region in the Twentieth Century*. Cornell University Press, 2005.
The Dublin Statement and Report of the Conference. International Conference on Water and the Environment: Development Issues for the 21st Century, Dublin, Ireland, 26–31 Jan. 1992. https://www.ircwash.org/sites/default/files/71-ICWE92-9739.pdf. Accessed 17 Aug. 2020.
Durfee, J. R. *Reminiscences of Carbondale, Dundaff, and Providence, Forty Years Past*. Miller's Bible Publishing, 1875.
"Duryea the Scene of Wild Tumult." *Scranton Tribune*, 15 Aug. 1902: 3.
"Dying from Neglect." *Scranton Republican*, 27 Aug. 1892: 5.
"Dynamite Blast Begins Work on Huge Stillwater Reservoir." *Scranton Times*, 14 May 1958: 3, 31.
"Dynamiting at the National." *Scranton Tribune*, 1 Sept. 1902: 3.
"Early Railroad Fight." *Pittston Gazette*, 31 July 1908: 6.
"Eastern Coal and Coke Notes." *Black Diamond* 23, no. 15 (1899): 416.
Editorial. *Scranton Republican*, 13 Aug. 1902: 4.
"Editorial Notes." *Scranton Times*, 3 Mar. 1902: 4.
"Editorial Notes." *Scranton Times*, 9 Aug. 1902: 4.
Eilperin, Juliet, and Dennis Brady. "EPA Rule Requires Raw Research Data." *Washington Post*, 6 Jan. 2021.
Ellis, Erle C. *Anthropocene: A Very Short Introduction*. Oxford University Press, 2018.
Emmett, Robert S., and David E. Nye. *The Environmental Humanities: A Critical Introduction*. MIT Press, 2017.
"Encroaching on the River Banks." *Scranton Republican*, 20 May 1902: 2.
"Encroachment upon River Channel a Menace to All Living on the Low Lands." *Scranton Times*, 17 Oct. 1903: 1, 7.
Engel, Andrew D., and Clayton J. Ruminski. "Pennsylvania Power & Light Company Records." Finding aid historical note. Hagley Museum archives, 2014 and 2017. https://findingaids.hagley.org/repositories/3/resources/932. Accessed 9 Feb. 2021.
"Ensure Funds to Continue Recovery Work." Editorial. *Scranton Times-Tribune*, 25 Jan. 2020: A10.
"Environmental Vandalism." Editorial. *Scranton Times-Tribune*, 17 Aug. 2020: A8.
EPA (Environmental Protection Agency). "Clean Power Plan for Existing Power Plants: Regulatory Actions." 22 Dec. 2015. https://archive.epa.gov/epa/cleanpowerplan/clean-power-plan-existing-power-plants-regulatory-actions.html. Accessed 7 May 2024.
———. "EPA Implements Court Decision Overturning Restrictive Trump-Era Rule; Reaffirms Commitment to Use Best Available Science." 26 May 2021. https://www.epa.gov/newsreleases/epa-implements-court-decision-overturning-restrictive-trump-era-rule-reaffirms. Accessed 7 May 2024.

EPCAMR (Eastern Pennsylvania Coalition for Abandoned Mine Reclamation). "3D Mine Pool Mapping." See minute 3:33 in film *3D Mine Pool Model of the Northern Anthracite Coal Field.* https://epcamr.org/home/current-initiatives/technical-assistance/watershed-assessment/mine-pool-mapping-initiative/. Accessed 31 Mar. 2021.

"Erie Lackawanna Files for Bankruptcy." *Philadelphia Inquirer*, 27 June 1972: 16.

Etiope, G., G. Papatheodorou, D. Christodoulou, M. Geraga, and P. Favali. "The Geological Links of the Ancient Delphic Oracle (Greece): A Reappraisal of Natural Gas Occurrence and Origin." *Geology* 34, no. 10 (Oct. 2006): 821–824. https://doi.org/10.1130/G22824.1.

Evans, Robley D. "Reserve Our Anthracite for Our Navy." *North American Review* 84, no. 608 (1907): 246–253.

"An Example to Follow." *Scranton Republican*, 6 Feb. 1901: 4.

"Excessive Heat Causes Close to 200 Deaths in New York." *Scranton Times*, 3 Aug. 1917: 20.

"Extension of the Strike." *Scranton Tribune*, 13 Nov. 1900: 2.

"Extracts." *Pittston Gazette*, 13 July 1855: 2.

"Fact Check: Trump's State of the Union Address." NPR. 30 Jan. 2018. https://www.npr.org/2018/01/30/580378279/trumps-state-of-the-union-address-annotated. Accessed 22 July 2020.

Falchek, David. "Old Forge Borehole Draining Mines 50 Years." *Scranton Times-Tribune*, 26 Dec. 2012: A1, A7.

"Famous Freak of Nature Now Garbage Dump." *Wilkes-Barre Evening News*, 23 Oct. 1926: 3.

Farrell, Michael A. "The Use of Anthracite Coal as a Filter Medium." *Journal (American Water Works Association)* 25, no. 5 (May 1933): 718–724.

"A Fearful Fate." *Pottsville Republican*, 5 Feb. 1891: 3.

Feldman, Abraham. "Thoughts on Thales." *Classical Journal* 41, no. 1 (Oct. 1945): 4–6.

Felegy, E. W., L. H. Johnson, and J. Westfield. *Acid Mine Water in the Anthracite Region of Pennsylvania*. Technical Paper 710, U.S. Department of the Interior, Bureau of Mines. GPO, 1948.

"A Few Collieries Resume." *Scranton Republican*, 14 Oct. 1902: 7.

"A Fiendish Deed." *Scranton Republican*, 20 Aug. 1902: 5.

"Fight over the Lehigh." *Scranton Tribune*, 29 Oct. 1895: 3.

Finch, Robert, and John Elder, eds. *Norton Book of Nature Writing*. College ed. Norton, 2002.

Fineout, Gary, and Mark Sherman. "Fla., Ga. Water Fight Awaits Supreme Court." *Scranton Times-Tribune*, 8 Jan. 2018: B10.

Finkel, Madelon L. *Pipeline Politics: Assessing the Benefits and Harms of Energy Policy*. Praeger, 2018.

"Fire Destroys Small Washery." *Scranton Times*, 18 Sept. 1902: 5.

"First Official Step." *Scranton Times*, 15 Oct. 1903: 3, 5.

Fishback, Price V., and Shawn Everett Kantor. "The Adoption of Workers' Compensation in the United States, 1900–1930." *Journal of Law and Economics* 41, no. 2 (Oct. 1998): 305–341.

"Fish Commission Delays Recovery on Loss for Fish." *Wilkes-Barre Times Leader*, 24 Oct. 1961: 13.

Flannery, Joseph X. "Flood Death Toll May Hit 79." *Scrantonian*, 21 Aug. 1955: 1, 20.
"Flood Closes Schools and Mines in Duryea; Ohio Valley in Danger." *Scranton Republican*, 23 Mar. 1936: 1, 3.
"Flood Control Omitted Mines." *Scranton Republican*, 26 Mar. 1936: 1, 5.
"Flood in the Schooley Mine." *Scranton Tribune*, 26 May 1899: 7.
"Flood Stops Traffic." *Lancaster Examiner*, 14 Oct. 1903: 2, 7.
"Flushing of Culm into River from the Many Washeries Obstructs the Channel." *Scranton Times*, 19 Oct. 1903: 4.
Folsom, Burton. *Urban Capitalists: Entrepreneurs and City Growth in Pennsylvania's Lackawanna and Lehigh Regions, 1800–1920*. 2nd ed. University of Scranton Press, 2001. Rpt., Johns Hopkins University Press, 1981.
Folsom, Ed. "'I Have Been a Long Time in a Strange Country': W. S. Merwin and America." In *W. S. Merwin: Essays on the Poetry*, edited by Cary Nelson and Ed Folsom, 224–249. University of Illinois Press, 1987.
Fontana, Tom. "Transload Facility Services to Natural Gas Industry Increases." Tri-County Independent, 1 Feb. 2012. https://www.tricountyindependent.com/story/business/2012/02/01/transload-facility-services-to-natural/63630518007/. Accessed 6 May 2024.
"Forest City Coal and the Railroads." *Forest City News*, 1 Aug. 2018: 2.
Foster, Thomas J. *The Coal and Metal Miners' Pocketbook of Principles, Rules, Formulas, and Tables*. 7th ed. International Textbook Company, 1902.
———. *Coal Miners' Pocketbook*. 11th ed. McGraw-Hill, 1916. HathiTrust. https://babel.hathitrust.org/cgi/pt?id=uc2.ark:/13960/t4rj4b364&view=1up&seq=9&skin=2021. Accessed 10 May 2024.
"Four Companies after It." *Scranton Republican*, 16 Oct. 1895: 8.
"Four Strikers Shot in Riot at Duryea." *Scranton Times*, 14 Aug. 1902: 1.
Fowler, Lara. "New Special Master Finds for Georgia in Most Recent Round of Water Dispute." *SCOTUSblog*, 9 Jan. 2020. https://www.scotusblog.com/2020/01/new-special-master-finds-for-georgia-in-most-recent-round-of-water-dispute/. Accessed 9 May 2024.
"A Freak of the Glacial Period." *Bloomsburg Columbian*, 28 Mar. 1884: 2.
Freese, Barbara. *Coal: A Human History*. Perseus, 2003.
Freyfogle, Eric T. *The Land We Share: Private Property and the Common Good*. Island Press, 2003.
Friedman, Lisa. "E.P.A. Relaxes Rules Limiting Toxic Waste from Coal Plants." *New York Times*, 31 Aug. 2020. https://www.nytimes.com/2020/08/31//climate/trump-coal-plants.html. Accessed 29 Oct. 2023.
Frost, Robert. "For Once, Then, Something." In *The Poetry of Robert Frost*, edited by Edward Connery Lathem, 225. Henry Holt, 1979.
———. "The Pasture." In *The Poetry of Robert Frost*, edited by Edward Connery Lathem, 1. Henry Holt, 1979.
———. "The Prerequisites: A Preface." In *Robert Frost: Poetry and Prose*, edited by Edward Connery Lathem and Lawrance Thompson, 416–478. Henry Holt, 1984.
Fuller, Thomas. "Timber Company Tells California Town, Go Find Your Own Water." *New York Times*, 1 Oct. 2016. https://www.nytimes.com/2016/10/02/us/california-drought-weed-mount-shasta.html?searchResultPosition=1. Accessed 13 Mar. 2021.

"Funeral of John Jermyn." *Scranton Republican*, 2 June 1902: 6.

"A Further Supplement to 'An Act to Improve the Navigation of the River Lackawaxen." 5 Apr. 1826. In *Acts of the General Assembly of the Commonwealth of Pennsylvania*, 204–206. Cameron & Krause, 1826.

"Future of the City Demands That Action Be Taken to Stop Damage by River Floods." *Scranton Times*, 14 Oct. 1903: 5.

Gallagher, William J. "Hurricane Diane: Last Installment." Railfan.net. 8 July 1998. http://www.railfan.net/lists/erielack-digest/199807/msg00127.html. Accessed 28 July 2020.

Gardner, John. *On Moral Fiction*. New York: Basic Books, 1978.

"Gas and Water Co. Wins." *Scranton Republican*, 16 Dec. 1896: 7.

"Gas Explosion over Mined Area Brings Death and Injuries." *Scranton Times*, 20 Feb. 1924: 1, 13.

"Gas Line Tour Included Pipe Installation Work." *Forest City News*, 27 Sept. 2017: 2.

"A Gas Range." Advertisement. *Scranton Tribune*, 30 May 1902: 7.

"Gathering Winter Coal Supply." *Scranton Times*, 19 Oct. 1903: 8.

Gaydos, Ellyn McCormack. "Writing the Soul of a Place: An Interview with Jennifer Haigh." *Columbia Journal*, 20 Feb. 2019. http://columbiajournal.org/writing-the-soul-of-a-place-an-interview-with-jennifer-haigh/. Accessed 24 Oct. 2020.

General Accounting Office. "Carbon Capture and Storage: Actions Needed to Improve DOE Management of Demonstration Projects." 20 Dec. 2021. https://www.gao.gov/products/gao-22-105111. Accessed 7 May 2024.

"General Gobin in the City En Route to Olyphant." *Scranton Times*, 23 Sept. 1902: 1, 7.

Genesis. In *Catholic Study Bible*, edited by Donald Senior, Mary Ann Getty, Carroll Stuhlmueller, and John J. Collins, 2–62. Oxford University Press, 1990.

Gibbons, Brendan. "Long Time Burning: Mine Fires Widespread, Hard to Douse." *Scranton Times-Tribune*, 31 May 2014: A8.

Gibbons, Brendan, and Elizabeth Skrapits. "DEP to Fill Mine Voids in Luzerne." *Scranton Times-Tribune*, 31 Oct. 2013: A6.

"Glacial Pothole Now Public Property." *Pittston Gazette*, 18 Oct. 1940: 4.

"Glacial Pot-Holes." *Wilkes-Barre Record*, 30 Apr. 1897: 5.

Gleick, Peter H. "An Introduction to Global Fresh Water Issues." In *Water in Crisis: A Guide to the World's Fresh Water Resources*, edited by Peter H. Gleick, 3–12. Oxford University Press, 1993.

———. "Water and Energy." In *Water in Crisis: A Guide to the World's Fresh Water Resources*, edited by Peter H. Gleick, 67–79. Oxford University Press, 1993.

Gleick, Peter H., and Meena Palaniappan. "Peak Water Limits to Freshwater Withdrawal and Use." *Proceedings of the National Academy of Sciences of the United States of America* 107, no. 25 (2010): 11155–11162.

"Glen Alden Again Ordered to Keep Acid Water from River." *Hazleton Standard-Speaker*, 2 Dec. 1961: 2.

Golay, Faith. "Taylor Residents to Air Demands." *Scranton Times*, 22 July 1999: 26.

Gold, Russell. *The Boom: How Fracking Ignited the American Energy Revolution and Changed the World*. Simon & Schuster, 2014.

Goodell, Jeff. *The Water Will Come: Rising Seas, Sinking Cities, and the Remaking of the Civilized World*. Little, Brown, 2017.

Goodrich, Phineas. *History of Wayne County*. Haines and Beardsley, 1880.

Graves, Robert. *The Greek Myths*. Vol. 1. Penguin, 1955.
"Great Damage in Path of the Flood." *Scranton Tribune*, 1 Mar. 1902: 1, 6.
"Great Freshet on the Lackawanna." *Hazard's Register* 6 (Dec. 1830): 358.
"Green Ridge." *Scranton Times*, 17 June 1902: 9.
Griffen, John. "Slush Problem in Anthracite Preparation." *Transactions of the American Institute of Mining and Metallurgical Engineers* 66 (1922): 514–534.
Griswold, Eliza. *Amity and Prosperity: One Family and the Fracturing of America*. Farrar, Straus, and Giroux, 2018.
Grundfest, Jerry. "George Clymer: Philadelphia Revolutionary, 1739–1813." Ph.D. diss., Columbia University, 1973.
Haggerty, James. "Sew It Goes." *Scranton Sunday Times*, 4 Aug. 2013: H1, H6.
Haigh, Jennifer. *Heat and Light*. Ecco, 2016.
Hall, Elizabeth Armstrong. "If Looms Could Speak: The Story of Pennsylvania's Silk Industry." *Pennsylvania Heritage* 32, no. 3 (Summer 2006): 26–35. http://paheritage.wpengine.com/article/if-looms-could-speak-story-pennsylvanias-silk-industry/. Accessed 11 Mar. 2021.
Halpin, Bill. "Phoebe Snow Travels West for Final Time Today as an Era Ends." *Scrantonian*, 27 Nov. 1966: section II, 23, 26.
Hanson, Amy Beth, and Matthew Brown. "Turbulent Air." *Scranton Times-Tribune*, 2 Apr. 2020: B5.
Haraway, Donna. *Staying with the Trouble: Making Kin in the Chthulucene*. Duke University Press, 2016.
Harrison, Michael J., Stephen Marshak, and John H. McBride. "The Lackawanna Synclinorium, Pennsylvania: A Salt-Collapse Structure, Partially Modified by Thin-Skinned Folding." *GSA Bulletin* 116, no. 11/12 (Nov./Dec. 2004): 1499–1514. https://doi.org/10.1130/B25400.1.
Hasco, Linda. "These Are the 10 Biggest Floods the Susquehanna River Has Ever Seen." *Harrisburg Patriot-News*, PennLive, 25 July 2018. https://www.pennlive.com/news/erry-2018/07/4545db0ff19996/these-are-the-10-biggest-flood.html. Accessed 9 May 2024.
Hauter, Wenonah. *Frackopoly: The Battle for the Future of Energy and the Environment*. New Press, 2016.
Healey, Richard G. *The Pennsylvania Anthracite Coal Industry, 1860–1902*. University of Scranton Press, 2007.
"Health Officials Offer Council Sewage Report." *Scranton Times*, 9 May 1962: 3.
"Hearing Was Continued." *Scranton Times*, 26 June 1903: 1.
"Heavy Damage Caused throughout Region by Raging Flood Waters." *Scranton Times*, 5 Apr. 1947: 3, 7.
Heise, Ursula K. *Sense of Place and Sense of Planet: The Environmental Imagination of the Global*. Oxford University Press, 2008.
"He Loves That Dirty Water." Editorial. *Scranton Times-Tribune*, 24 Jan. 2020: A8.
Henry, Matthew S. "Extractive Fictions and Postextraction Futurisms: Energy and Environmental Injustice in Appalachia." *Environmental Humanities* 11, no. 2 (Nov. 2019): 402–426. https://doi.org/10.1215/22011919-7754534.
"Henry M. Stanley." Advertisement. *Scranton Republican*, 1 Dec. 1886: 2.
"Hetch-Hetchy Dam Dedicated." *Los Angeles Sunday Times*, 8 July 1923: 2.
"History of Strike." *Scranton Republican*, 22 Oct. 1902: 5.

Hitchcock, Frederick L. *History of Scranton and Its People*. Vol. 1. Lewis Historical Publishing, 1914. https://archive.org/details/historyofscranto01hitc/page/82/mode/2up. Accessed 14 Mar. 2021.

Hix, H. L. *Understanding W. S. Merwin*. University of South Carolina Press, 1997.

Hmiel, Benjamin, et al. "Preindustrial CH4 Indicates Greater Anthropogenic Fossil CH4 Emissions." *Nature* 578, no. 409–412 (2020). https://www.nature.com/articles/s41586-020-1991-8. Accessed 23 Mar. 2021.

Hollister, Horace. *History of the Lackawanna Valley*. 2nd ed. C. A. Alvord, 1869.

———. "Samuel Meredith." Letter to the editor. *Historical Record* 1, no. 12 (August 1887): 207.

Hollowell, Jerrald R., and Harry E. Koester. *Ground-Water Resources of Lackawanna County, Pennsylvania*. 4th Series, Water Resource Report 41. U.S. Geological Survey and Pennsylvania Geological Survey, 1975.

Homer. *The Odyssey*. Translated by Walter Shewring. Oxford University Press, 1980.

"Honesdale." *Scranton Tribune*, 11 Sept. 1895: 8.

"Honesdale and Hawley Hard Hit; Property Damage High; Carbondale Loss Enormous." *Scranton Times*, 25 May 1942: 3, 20.

Hopton, Sarah Beth, and Zackary Vernon. "From the Sublime to Destruction: An Introduction to 'Overburden' and Interview with Documentarian Chad A. Stevens." *Cold Mountain Review* 44, no. 2 (Spring 2016). https://www.coldmountainreview.org/issues/spring-2016/from-the-sublime-to-destruction-an-introduction-to-overburden-and-interview-with-documentarian-chad-a-stevens-by-sarah-beth-hopton-and-zackary-vernon. Accessed 27 July 2020.

"How Army Engineers' Map Shows Need for Flood Control in Wyoming Valley." *Wilkes-Barre Record*, 19 Mar. 1936: 3.

"How Does Your State Make Electricity?" *New York Times*, 24 Dec. 2018, updated in 2020. https://www.nytimes.com/interactive/2018/12/24/climate/how-electricity-generation-changed-in-your-state.html?action=click&linked=google. Accessed 9 May 2024.

Hoy, Suellen. *Chasing Dirt: The American Pursuit of Cleanliness*. Oxford University Press, 1995.

Hudson Coal Company. *The Story of Anthracite*. Hudson Coal Company, 1932.

Hughes, Robert. "Reimagine Future." Letter. *Scranton Sunday Times*, 18 Oct. 2020: C3.

Hughes, Robert E., and Roger J. Hornberger. "Introduction to the Anthracite Coal Region." In *Mine Water Resources of the Anthracite Coal Fields of Eastern Pennsylvania*, 6–20. N.d. [2011?]. Eastern PA Coalition for Abandoned Mine Reclamation. http://www.epcamr.org/storage/projects/MinePoolMapping/Mine_Water_Resources_of_the_Anthracite_Coal_Fields_-_Report.pdf. Accessed 7 May 2024.

Hughes, Robert E., Roger J. Hornberger, and Michael A. Hewitt. "The Development and Demise of Major Mining in the Northern Anthracite Coal Field." In *Mine Water Resources of the Anthracite Coal Fields of Eastern Pennsylvania*, 72–83. N.d. [2011?]. Eastern PA Coalition for Abandoned Mine Reclamation. http://www.epcamr.org/storage/projects/MinePoolMapping/Mine_Water_Resources_of_the_Anthracite_Coal_Fields_-_Report.pdf. Accessed 7 May 2024.

Hughes, Robert E., Roger J. Hornberger, Caroline M. Loop, Keith B. C. Brady, and Nathan A. Houtz. "Geology of the Anthracite Coal Region." In *Mine Water Resources of the Anthracite Coal Fields of Eastern Pennsylvania*, 21–48. N.d. [2011?]. Eastern

PA Coalition for Abandoned Mine Reclamation. http://www.epcamr.org/storage/projects/MinePoolMapping/Mine_Water_Resources_of_the_Anthracite_Coal_Fields_-_Report.pdf. Accessed 7 May 2024.

Hughes, Robert E., Roger J. Hornberger, David L. Williams, Daniel K. Koury, and Keith A. Laslow. "Colliery Development in the Anthracite Coal Fields." In *Mine Water Resources of the Anthracite Coal Fields of Eastern Pennsylvania*, 49–60. N.d. [2011?]. Eastern PA Coalition for Abandoned Mine Reclamation. http://www.epcamr.org/storage/projects/MinePoolMapping/Mine_Water_Resources_of_the_Anthracite_Coal_Fields_-_Report.pdf. Accessed 7 May 2024.

Hunt, Bruce J. *Pursuing Power and Light: Technology and Physics from James Watt to Albert Einstein*. Johns Hopkins University Press, 2010.

"Hurrah for Fourth; Big Celebration." *Scranton Times*, 3 July 1909: 1.

"An Illustrious Family—The Pedigree of the Merediths." *Wayne County Herald*, 6 Feb. 1879: 2.

"In Common Council." *Scranton Republican*, 18 Nov 1891: 5.

"Increased Water Supply." *Scranton Republican*, 18 Oct. 1895: 5.

Ineson, Frank A., and Miles J. Ferree. *Anthracite Forest Region: A Problem Area*. GPO, 1948.

Ingraffea, Anthony R., Paul A. Wawrzynek, Renee Santoro, and Martin Wells. "Reported Methane Emissions from Active Oil and Gas Wells in Pennsylvania, 2014–2018." *Environmental Science and Technology* 54, no. 9 (2020): 5783–5789. https://doi.org/10.1021/acs.est.0c00863.

Inners, Jon D., James A. LaRegina, and Alex Chamberlin. "'King Coal' on the Mountain: Geology, Mining History, and Engineering of the Hazleton Shaft Colliery, Northeastern Pennsylvania." Abstract of poster paper presentation at the 35th Annual Meeting of the Northeastern Section, Geological Society of America, New Brunswick, NJ, 15 Mar. 2000. In Geological Society of America. *Abstracts with Programs*, 32, no. 1: 26.

Intergovernmental Panel on Climate Change. "Summary for Policy Makers." Climate Change 2021: The Physical Science Basis. https://www.ipcc.ch/report/ar6/wg1/downloads/report/IPCC_AR6_WGI_SPM.pdf: 1–31. Accessed 25 Sept. 2021.

"Internal Improvement." *Wyoming Herald*, 11 Nov. 1825: 3.

"Internal Improvement." *Wyoming Herald*, 23 Dec. 1825: 3.

"Internal Improvement." *Wyoming Herald*, 13 Jan. 1826: 2.

International Correspondence Schools. "Manufacture of Gas." Part 1, no. 54. In *Manufacture of Cement; Manufacture of Paper; Manufacture of Sugar; Petroleum and Products; Manufacture of Gas*, 1–42. International Textbook, 1902.

———. "Manufacture of Gas." Part 2, no. 55. In *Manufacture of Cement; Manufacture of Paper; Manufacture of Sugar; Petroleum and Products; Manufacture of Gas*, 1–42. International Textbook, 1902.

"Is Fracking Linked to Oklahoma Earthquakes?" *Scranton Sunday Times*, 6 May 2018: A16.

"Is Violence Near?" *Scranton Republican*, 5 June 1902: 1.

Jablonka, Ivan. *History Is a Contemporary Literature*. Translated by Nathan J. Bracher. Cornell University Press, 2018.

James, Tom. "Water Hazard." *Scranton Times-Tribune*, 25 June 2018: B7.

"John Jermyn's Busy Life Is at an End." *Scranton Tribune*, 30 May 1902: 6.

Johnson, Bob. *Carbon Nation: Fossil Fuels in the Making of American Culture*. University Press of Kansas, 2014.

Johnson, Emily [Caspar Day]. "Good Manners and the Water Company." *Century Magazine*, Nov. 1908: 57–66.

Jones, Christopher F. *Routes of Power: Energy and Modern America*. Harvard University Press, 2014.

"The Judicial History of the Anthracite Monopoly." *Yale Law Journal* 41, no. 3 (Jan. 1932): 439–444. https://doi.org/10.2307/791007.

Kaiser, Anna Jean, and Marcelo Silva De Sousa. "Dam Collapse Proves Deadly." *Scranton Times-Tribune*, 26 Jan. 2019: A8.

Kashmer, Michael A. "River Recollections: Scranton's Plot Section." *History Bytes*, Mar.–Apr. 2020: 7–11.

Keller, C. Brenhin, et al. "Neoproterozoic Glacial Origin of the Great Unconformity." *Proceedings of the National Academy of Sciences* 116, no. 4 (31 Dec. 2018): 1136–1145. https://doi.org/10.1073/pnas.1804350116.

Kelley, Paul. "The River." In *Coalseam: Poems from the Anthracite Region*, 2nd ed., edited by Karen Blomain, 38–39. University of Scranton Press, 1996.

Kelo et al. v. City of New London. 23 June 2005. Legal Information Institute, Cornell Law School, Cornell University. https://www.law.cornell.edu/supct/html/04-108.ZS.html. Accessed 21 Mar. 2012.

Kessler, Glenn. "Biden's Claim That His 'Great-Granpop' Was a Coal Miner." *Washington Post*, 20 May 2021. https://www.washingtonpost.com/politics/2021/05/20/bidens-claim-that-his-great-grandpop-was-coal-miner/. Accessed 10 May 2024.

Kneeland, Timothy W. *Playing Politics with Natural Disaster: Hurricane Agnes, the 1972 Election, and the Origins of FEMA*. Cornell University Press, 2020.

Knickmeyer, Ellen. "Trump Lifts Waterways Protections." *Scranton Times-Tribune*, 24 Jan. 2020: B9.

Koenig, David. "U.S. 'Likely' Has Taken Over as World's Top Oil Producer." *Scranton Times-Tribune*, 13 Sept. 2018: B9.

Korb, Michael C. "The Conowingo Tunnel and the Anthracite Mine Flood-Control Project: A Historical Perspective on a 'Solution' to the Anthracite Mine Drainage Problem." Paper presented at the 13th Annual PA Conference on Abandoned Mine Reclamation and Coal Mine Heritage, Hazleton, PA, August 2011. https://files.dep.state.pa.us/Mining/Abandoned%20Mine% 20Reclamation/AbandonedMinePortalFiles/Publications/AMLRelatedTechnicalPapers/Conowingo_Tunnel-2011.pdf. Accessed 15 Mar. 2021.

Korson, George, ed. *Songs and Ballads of the Anthracite Miner*. F. H. Hitchcock, 1927.

Kowalski, Paul. "River Recollections: Olyphant." *History Bytes*, Mar.–Apr. 2020: 12–13.

Krawczeniuk, Borys. "Candidates Dwell on Area's Long-Gone Coal Legacy." *Scranton Times-Tribune*, 10 Sept. 2016: A11.

———. "Freight Railroad on the Right Track." *Scranton Sunday Times*, 7 Jan. 2018: C4, C6.

———. "Local Railroad Traffic Up Again." *Scranton Sunday Times*, 9 Dec. 2018: C4, C5.

———. "A Record-Breaking Year." *Scranton Times-Tribune*, 14 Jan. 2019: A3.

Kummer, Frank. "Here's What Climate Change Will Look Like in Northeast." *Scranton Times-Tribune*, 1 Dec. 2018: B9.

"The Lace Factory." *Scranton Republican*, 3 May 1892: 5.

"Lackawanna May Build Huge Dam in Pike County." *Scranton Times*, 20 Jan. 1921: 14.

"Lackawanna Negotiating for the Purchase of Big Watershed Near Hawley." *Scranton Times*, 13 June 1922: 1.

"Lackawanna Railroad." Advertisement. *New York Times*, 14 Apr. 1909: 9.

Lackawanna Railroad. *The Story of Phoebe Snow*. Lackawanna Railroad, n.d. [1943?]. https://hoboken.pastperfectonline.com/archive/F4E272F2-9E68-4CCC-B920-420695298139. Accessed 31 Mar. 2021.

"Lackawanna River Pours into Glen Alden Mines." *Carbondale Leader*, 8 Apr. 1924: 1, 4.

"Lake Elmhurst or Dam-Site—Which?" *Scranton Weekly Republican*, 20 Nov. 1889: 5.

Lamm, Mariah, Lisa Hallowell, Abel Russ, and Tom Pelton. *Sewage Overflows in Pennsylvania's Capital*. Environmental Integrity Project. 22 Aug. 2019. https://www.environmentalintegrity.org/wp-content/uploads/2019/08/PA-Sewage-Report-Final.pdf. Accessed 19 July 2020.

"Large Reservoir Burst." *Scranton Tribune*, 11 Oct. 1895: 5.

"Largest Lake in Pennsylvania to Be Created for Purpose of Generating Electric Power." *Scranton Truth*, 18 June 1910: 1.

Lassiter, Luke Eric. "The Elk River Spill: On Water and Trust." In *I'm Afraid of That Water: A Collaborative Ethnography of a West Virginia Water Crisis*, edited by Luke Eric Lassiter, Brian A. Hoey, and Elizabeth Campbell, 17–48. West Virginia University Press, 2020.

Lauver, Fred J. "A Walk through the Rise and Fall of Anthracite Might." *Pennsylvania Heritage Magazine* 27, no. 1 (Winter 2001): n.p. http://www.phmc.state.pa.us/portal/communities/pa-heritage/walk-through-rise-fall-anthracite-might.html. Accessed 16 Mar. 2021.

Laws of Pennsylvania Genealogical Data. Compiled by Vi P. Limric. 1998. USGenWeb Archives. http://files.usgwarchives.net/pa/1pa/xmisc/1826laws.txt. Accessed 1 Dec. 2020.

LeCain, Timothy J. *The Matter of History: How Things Create the Past*. Cambridge University Press, 2017.

Lee, Richard. "The Recovery of Anthracite from Culm Banks." *Engineering and Mining Journal* 85 (4 Apr. 1908): 720–722.

"Legal." *Scranton Tribune*, 6 Feb. 1899: 2.

"The Lehigh River Dams." *Scranton Republican*, 7 May 1896: 6.

"Lehigh River Water." *Scranton Republican*, 11 Dec. 1895: 5.

Leighton, Marshall Ora. *Quality of Water in the Susquehanna River Drainage Basin*. GPO, 1904.

LeMenager, Stephanie. *Living Oil: Petroleum Culture in the American Century*. Oxford University Press, 2014.

Lenton, Timothy M., Johan Rockstrom, Owen Gaffney, Stefan Rahmstorf, Katherine Richardson, Will Steffan, and Hans Joachim Schnellnhuber. "Climate Tipping Points— Too Risky to Bet Against." *Nature* 575 (2019): 592–595. https://doi.org/10.1038/d41586-019-03595-0.

LeRoy, Edwin D. *The Delaware and Hudson Canal: A History*. Wayne County [PA] Historical Society, 1950.

Lesnefsky, Frank Wilkes. "Millions Pumped into Dolph Mine Fire." *Scranton Times-Tribune*, 13 Feb. 2020: A1, A11.

———. "Wind Farm Plan Fans Concerns." *Scranton Times-Tribune*, 7 June 2019: A1, A4.

Leung, Rebecca. "A Toxic Cover-Up?" CBS News. *60 Minutes*, 1 Apr. 2004. https://www.cbsnews.com/news/a-toxic-cover-up/. Accessed 27 July 2020.

Levy, Marc. "Pennsylvania Orders Gas Well Plugged in Fight over Methane." *Scranton Times-Tribune*, 14 Jan. 2020: A5.

Lewis, William P. "The Life and Work of Col. L. A. Watres." *Lackawanna Historical Society Bulletin* 16, no. 2 (Apr. 1983): n.p.

Lieberman, Benjamin, and Elizabeth Gordon. *Climate Change in Human History: Prehistory to the Present.* Bloomsbury, 2018.

List of Lands Belonging to the Estate of S. Meredith, Susquehanna County. Archives of the Susquehanna County Historical Society, Montrose, PA.

"Local Gas Plant Furnishing Light to Big Clientele." *Pittston Gazette*, 21 Apr. 1927: 5.

Lockwood, Jim. "College Gets Funds to Fill Mine Void." *Scranton Times-Tribune*, 11 Apr. 2018: A3, A7.

———. "Upgrades Due to Antiquated Sewer Lines." *Scranton Times-Tribune*, 28 Aug. 2016: A8.

———. "The Water Is Fine: Lackawanna River Wins Pennsylvania River of Year Title through Vote." *Scranton Times-Tribune*, 30 Jan. 2020: A3, A5.

"Looking Back." *Forest City News*, 21 Dec. 2016: 4.

"Looking Back." *Forest City News*, 28 Feb. 2018: 6.

Lowenthal, Larry. *From the Coalfields to the Hudson: A History of the Delaware & Hudson Canal.* Purple Mountain Press, 1997.

Lund, Jay, Josue Medellin-Azuara, John Durand, and Kathleen Stone. "Lessons from California's 2012–2016 Drought." *Journal of Water Resource Planning and Management* 144, no. 10 (2018): 1–13. https://doi.org/10.1061/(ASCE)WR.1943-5452.0000984.

Lynch, Tom, Cheryll Glotfelty, and Karla Armbruster. Introduction to *The Bioregional Imagination: Literature, Ecology, and Place*, edited by Tom Lynch, Cheryll Glotelty, and Karla Armbruster, 1–29. University of Georgia Press, 2012.

Lyon, Paul. "River Projects Need Bond $$." *Scranton Times*, 27 June 1994: 3, 4.

MacLeod, Janine. "Holding Water in Times of Hydrophobia." In *Petrocultures: Oil, Politics, Culture*, edited by Sheena Wilson, Adam Carlson, and Imre Szeman, 264–286. McGill-Queen's University Press, 2017.

Magdy, Samy. "A Disaster That Was Waiting to Happen." *Scranton Times-Tribune*, 19 Sept. 2023: B10.

Malm, Andreas. *Fossil Capital: The Rise of Steam Power and the Roots of Global Warming.* Verso, 2016.

Manno, Jack P., and Stephen B. Balogh. "The Biophysical: The Decline in Energy Returned on Energy Invested, Net Energy, and Marginal Benefits." In *Ending the Fossil Fuel Era*, edited by Thomas Princen, Jack P. Manno, and Pamela L. Martin, 37–52. MIT Press, 2015.

"Many Washeries Being Operated by Imported Men." *Scranton Times*, 20 June 1902: 5.

Marcus, Lew. "State Grapples with Problem of River 'Mess.'" *Scrantonian*, 19 Aug. 1979: 23.

Martin, Dennis. "The Authors Carnival of 1886." *LHS Journal* 43, no. 2 (Spring 2013): 1, 7.

Maslin, Mark. *Climate Change: A Very Short Introduction*. 3rd ed. Oxford University Press, 2014.

Mathews, Alfred. *History of Wayne, Pike and Monroe Counties, Pennsylvania*. R. T. Peck, 1886.

Matthews, Christopher M. "The Next Big Bet in Fracking: Water." *Wall Street Journal*, 22 Aug. 2018. https://www.wsj.com/articles/the-next-big-bet-in-fracking-water-1534930200. Accessed 23 Mar. 2021.

Matthews, Jennifer. "Treated Hydraulic Fracturing Wastewater May Pollute Area Water Sources for Years." *Penn State News*, 13 July 2017. https://news.psu.edu/story/474649/2017/07/13/research/treated-hydraulic-fracturing-wastewater-may-pollute-area-water. Accessed 14 Oct. 2020.

Maxey, David W. "Of Castles in Stockport and Other Strictures: Samuel Preston's Contentious Agency for Henry Drinker." *Pennsylvania Magazine of History and Biography* 100 (July 1986): 413–446.

"May Go to Shenandoah." *Scranton Republican*, 24 Sept. 1902: 7.

Maykuth, Andrew. "Fracking Suit Sent Back Down." *Philadelphia Inquirer*, 4 July 2018: A8.

McConnell, Steve. "Gas Pipeline Work to Begin." *Scranton Times-Tribune*, 10 June 2010: D1, D2.

McGlashen, Andy. "Biden Administration Restores Migratory Bird Treaty Act Protections." *Audubon Magazine*, 29 Sept. 2021. https://www.audubon.org/news/biden-administration-restores-migratory-bird-treaty-act-protections. Accessed 24 Oct. 2023.

McGraw, Seamus. *The End of Country*. Random House, 2011.

McGurl, Bernard. *The Lackawanna River Guide*. 2nd ed. Lackawanna River Corridor Association, 2002.

———. "Lackawanna River Stormwater Agency Needed." *Sunday Times*, 12 Aug. 2018: C2.

———. "Lackawanna River: Watershed Tour." Bus trip. 12 August 2017.

———. "Saving River Requires Citizen Involvement." *Times-Tribune*, 19 June 2016: D5.

McGurl, Bernard, Robert E. Hughes, and Michael Hewitt. *Lower Lackawanna River Watershed Restoration and Assessment Plan Report*. Lackawanna River Corridor Association, 2012. https://img1.wsimg.com/blobby/go/35928d34-4fb2-4795-b916-4e3efac60feb/downloads/LLWRAP_Final.pdf?ver=1605140549621. Accessed 15 Mar. 2021.

McGurl, Bernard, Arthur Popp, Deilsie Heath Kulesa, and Gail Puente. *Lackawanna River Watershed Conservation Plan*. Lackawanna River Corridor Association, 2001. http://elibrary.dcnr.pa.gov/GetDocument?docId=1736947&Doc Name=RiverRegistry52_LackawannaRiver.pdf. Accessed 12 Jan. 2021.

McKibben, Bill, ed. *American Earth: Environmental Writing since Thoreau*. Library of America, 2008.

McLean, Bethany. "The Next Financial Crisis Lurks Underground." *New York Times*, 1 Sept. 2018. https://www.nytimes.com/2018/09/01/opinion/the-next-financial-crisis-lurks-underground.html. Accessed 24 Mar. 2021.

McNeill, J. R. *Something New under the Sun: An Environmental History of the Twentieth-Century World*. W. W. Norton, 2000.

McNeill, J. R., and Peter Engelke. *The Great Acceleration: Environmental History of the Anthropocene since 1945*. Harvard University Press, 2014.

Melosi, Martin V. *The Sanitary City: Environmental Services in Urban America from Colonial Times to the Present*. Johns Hopkins University Press, 2000. Abridged ed., University of Pittsburgh Press, 2008.

"Men Tricked Here Under False Promises Refused to Work at No. 6 Washery." *Scranton Times*, 18 June 1902: 5.

Meredith, Thomas. Letter to Jason Torrey. 31 Dec. 1824. Meredith Papers. Archives of the Luzerne County Historical Society.

———. *Monopoly Is Tyranny; or, An Appeal to the People and Legislature, from the Oppression of the Delaware & Hudson Canal Company*. Pamphlet. S. Hamilton, 1830.

Merriam-Webster's Collegiate Dictionary. 11th ed. Merriam-Webster, 2003.

Merwin, W. S. "'Fact Has Two Faces': Interview." In *Regions of Memory: Uncollected Prose, 1949–82*, edited and with an introduction by Ed Folsom and Cary Nelson, 320–361. University of Illinois Press, 1987.

———. "Lackawanna." In *The Second Four Books of Poems*, 162–164. Copper Canyon Press, 1993.

———. "The Well." In *The Second Four Books of Poems*, 157–158. Copper Canyon Press, 1993.

"Met the Milk Dealers." *Scranton Republican*, 17 July 1896: 3.

Meylert, William, to Ada Meylert. Letter. 30 Dec. 1849. Secku Meylert binder. Sullivan County Historical Society.

Micek, John L. "Analysis: Trump Stumps for Lou Barletta in Pa., Attacking Media." *Harrisburg Patriot-News*, PennLive, 3 Aug. 2018, updated 29 Jan. 2019. https://www.pennlive.com/news/2018/08/ in_a_speech_riddled_with_false.html. Accessed 14 Mar. 2021.

Michalski, Stanley R. "The Jharia Mine Fire Control Technical Assistance Project: An Analysis." *International Journal of Coal Geology* 59 (2004): 83–90.

Miller, Donald L., and Richard E. Sharpless. *The Kingdom of Coal: Work, Enterprise, and Ethnic Communities in the Mine Fields*. Canal History and Technology Press, 1998. Rprt., University of Pennsylvania Press, 1985.

Milman, Oliver. "Renewables Surpass Coal in US Energy Generation for First Time in 130 Years." *The Guardian*, 3 June 2020.

"Mine Allowed to Fill Up." *Wilkes-Barre Record*, 21 June 1899: 5.

"Mine Drainage in Susquehanna Is Issue Inquiry into Killing of Fish." *Evening Times* [Sayre, PA], 19 Oct. 1961: 2.

"The Mine Horror." *Philadelphia Times*, 20 Dec. 1885: 1.

"Mines and Mining." *Carbondale Leader*, 18 Apr. 1893: 2.

"Mine Voids Use for Sewage Is Termed Illegal." *Scranton Times*, 27 Mar. 1963: 3, 28.

"Mine Workers Take a Hand." *Scranton Tribune*, 4 Jan. 1901: 2.

"A Mining Miracle." *Wilkes-Barre Evening Leader*, 24 Feb. 1891: 1.

"Mining of Anthracite." In *Mac's Directory and Handbook of Anthracite*, 19–21. Coal Information Bureau, 1937.

Mitchell, Timothy. *Carbon Democracy: Political Power in the Age of Oil*. Verso, 2013. Repr., 2011.

"Mitchell's Orders Generally Obeyed." *Scranton Times*, 2 June 1902: 1.

"Mitchell to be Here." *Scranton Republican*, 23 Aug. 1902: 5.

Molesky, Jason. "Gothic Toxicity and the Mysteries of Nondisclosure in American Hydrofracking Literature." *Modern Fiction Studies* 66, no. 1 (Spring 2020): 52–77. https://doi.org/10.1353/mfs.2020.0002.

"Money, Mine and Labor." *Wilkes-Barre Times*, 1 Oct. 1895: 4.

Morford, Mark P. O., and Robert J. Lenardon. *Classical Mythology*. 8th ed. Oxford University Press, 2007.

Morton, Timothy. *Dark Ecology: For a Logic of Future Coexistence*. Columbia University Press, 2016.

———. *Hyperobjects: Philosophy and Ecology after the End of the World*. University of Minnesota Press, 2013.

Mosley, Philip, ed. *Anthracite!: An Anthology of Pennsylvania Coal Region Plays*. University of Scranton Press, 2006.

"Mr. Jos. Jermyn Acts Promptly." *Scranton Republican*, 27 May 1902: 7.

"Mr. Scranton on the Water Rates." *Scranton Times*, 21 Nov. 1901: 3.

"Mr. Scranton Talks." *Scranton Republican*, 21 Dec. 1907: 5, 6.

Murphy, John. "Four Local Sewage Plants Called Chesapeake Bay's Worst Polluters." *Scranton Sunday Times*, 29 Dec. 1991: A1, A6.

Murphy, Thomas. *Jubilee History Commemorative of the Fiftieth Anniversary of the Creation of Lackawanna County, Pennsylvania*. Vols. 1–2. Historical Publishing Company, 1928.

NASA (National Aeronautics and Space Administration). "NASA Announces Summer 2023 Hottest on Record." Global Climate Change: Vital Signs of the Planet. 14 Sept. 2023. https://climate.nasa.gov/news/3282/nasa-announces-summer-2023-hottest-on-record/. Accessed 24 Oct. 2023.

National Integrated Drought Information System. "California." https://www.drought.gov/states/california. Accessed 19 Mar. 2021.

National Transportation Safety Board. "Natural Gas Pipeline Rupture and Fire." Investigation ID: DCA00MP009. https://www.ntsb.gov/investigations/Pages/DCA00MP009.aspx. Accessed 7 May 2024.

National Weather Service, National Oceanic and Atmospheric Administration. "Historical Floods: Susquehanna River at Wilkes-Barre, PA." 21 Jan. 2020. https://www.weather.gov/media/marfc/FloodClimo/MSU/WilkesBarre.pdf. Accessed 7 Jan. 2021.

———. "Local Month/Year Temperature & Precipitation Charts for Scranton, PA 2018." See chart "Year 2018." https://www.weather.gov/bgm/climatePlots2018AVP. Accessed 6 Aug. 2020.

———. "Local Month/Year Temperature & Precipitation Charts for Scranton, PA 2019." See chart "Year 2019." https://www.weather.gov/bgm/climatePlots2019AVP. Accessed 6 Aug. 2020.

———. "Normals for Scranton/Wilkes-Barre (Avoca), PA (1981–2010 Data)." https://www.weather.gov/bgm/climateAVPMonthlyNormals. Accessed 6 Aug. 2020.

Natural Gas Supply Association. "Processing Natural Gas." 25 Sept. 2013. http://naturalgas.org/naturalgas/processing-ng/. Accessed 23 Mar. 2021.

———. "The Transportation of Natural Gas." 20 Sept. 2013. http://naturalgas.org/naturalgas/transport/. Accessed 23 Mar. 2021.

"The Need of Public Baths." *Scranton Times*, 19 June 1899: 4.

"New Company 'Stands Ready and Has Offered to Reduce the Water Rates in the City.'" *Scranton Times*, 3 July 1903: 3, 5.
"New Dam Built." *Pottsville Miners Journal*, 26 Sept. 1895: 4.
"New Water Company for City of Scranton." *Scranton Times*, 26 May 1903: 5.
"New Water Project Formally Launched." *Carbondale Leader*, 27 May 1903: 5.
"19 Days Entombed." *Wilkes-Barre Daily News-Dealer*, 25 Feb. 1891: 1.
Nixon, Rob. *Slow Violence and the Environmentalism of the Poor*. Harvard University Press, 2011.
"No Hope and All Dead." *Wilkes-Barre Evening Leader*, 22 Dec. 1885: 1.
"No More Water Famines." *Scranton Republican*, 13 May 1896: 6.
"Non-Union Man Clubbed to Death; Foreman's Ear Cut Off by Pole." *Scranton Republican*, 26 Sept. 1902: 7.
Norris, R.V. "The Unwatering of Mines in the Anthracite Region." *Engineering Magazine* 34 (1907): 159–178.
"Notes of Tour of Scranton Lace Company, 10 March 1998." Scranton Lace Company file, 677 SCR 15 L116. Archives of the Lackawanna Historical Society.
"Nottingham Lace Manufacturers." *Scranton Tribune* [Greater Scranton edition], 27 Nov. 1897: 4.
Novak, Michael. *The Guns of Lattimer*. Basic Books, 1978.
"No Water-Rate War Says Mr. Jermyn." *Scranton Tribune*, 3 May 1902: 2.
"NTSB Releases Report on Explosion." *Santa Maria Times* [California], 27 Sept. 2011: A5.
O'Connell, Jon. "Driller Cabot Faces Charges." *Scranton Times-Tribune*, 16 June 2020: A1, A5.
———. "Hydroelectric Plan Takes a Step." *Scranton Times-Tribune*, 16 Oct. 2017: A3, A5.
———. "Power Plant Would Create Its Own Power." *Scranton Sunday Times*, 18 Oct. 2015: A 7.
———. "Vapor, Subsidence Concern Some Near Power Plant." *Scranton Times-Tribune*, 9 March 2018: A1, A11.
O'Connell, Jon, and Denise Allabaugh. "From NEPA, with Love." *Scranton Sunday Times*, 5 Aug. 2018: C4, C5.
Office of the Attorney General, Commonwealth of Pennsylvania. *Report 1 of the Forty-Third Statewide Investigating Grand Jury*. June 2020. https://www.attorneygeneral.gov/wp-content/uploads/2020/06/FINAL-fracking-report-w.responses-with-page-number-V2.pdf. Accessed 24 Mar. 2021.
"Officers Were Elected." *Scranton Tribune*, 29 Nov. 1899: 10.
Official Data Foundation, Alioth Finance. "$13,850 in 1817 → 2021 | Inflation Calculator." https://www.officialdata.org/us/inflation/1817?amount=13850. Accessed 20 Mar. 2021.
———."$27,000 in 1825 → 2024 | Inflation Calculator." https://www.officialdata.org/us/inflation/1825?amount=27000. Accessed 6 May 2024.
O'Hara, John. *Appointment in Samarra*. Harcourt, Brace, 1934.
"Old Gen. Humidity on Job; Weatherman Gives Hope of More Bearable Temperature." *Scranton Times*, 2 Aug. 1917: 1, 7.
Olson, Laura. "Report Mentions Interference in Pa." *Scranton Times-Tribune*, 19 Apr. 2019: A7.
"Olyphant." *Scranton Republican*, 18 Oct. 1900: 6.

Onishi, Norimitsu, and Somini Sengupta. "In South Africa, Facing 'Day Zero' with No Water." *New York Times*, 30 Jan. 2018: A1.
"On the Wing." *Black Diamond* 23, no. 19 (4 Nov. 1899): 523.
"Operators Feel That They Have the Advantage." *Scranton Tribune*, 3 June 1902: 1.
"Our Drinking Water." *Scranton Republican*, 12 Nov 1892: 5.
"Outbreaks at Four Places." *Scranton Tribune*, 21 May 1902: 5.
"Outlines Plans for Power Development." *Scranton Republican*, 19 July 1923: 2.
Parini, Jay. *Robert Frost: A Life*. Henry Holt, 1999.
Pearce, Fred. *When the Rivers Run Dry: Water—the Defining Crisis of the Twenty-First Century*. Beacon, 2006.
Pearson, Thomas W. *When the Hills Are Gone: Frac Sand Mining and the Struggle for Community*. University of Minnesota Press, 2017.
PennFuture and Conservation Voters of Pennsylvania. "We Should Never . . ." Advertisement. *Scranton Sunday Times*, 15 Mar. 2020: A13.
Pennsylvania Coal Company. *The Pennsylvania Story*. Pennsylvania Coal Company, n.d. [1939?].
"Pennsylvania Collieries Resume." *Carbondale Leader*, 5 Nov. 1895: 1.
Pennsylvania Department of Environmental Protection. "Climate Change in PA." https://www.depgis.state.pa.us/ClimateChange/index.html. Accessed 14 Jan. 2021.
Pennsylvania Department of Environmental Protection, Office of Active and Abandoned Mine Operations, Bureau of Abandoned Mine Reclamation. "Powderly Creek Northeast Underground Mine Fire." https://files.dep.state.pa.us/Mining/Abandoned%20Mine%20Reclamation/AbandonedMinePortalFiles/Accomplishments/OSM35_1520_102.1_Powderly.pdf. Accessed 13 Mar. 2021.
Pennsylvania Department of Internal Affairs. *Reports of the Inspectors of Mines of the Anthracite and Bituminous Coal Regions of Pennsylvania, 1891*. Edwin K. Myers, 1892.
———. *Reports of the Inspectors of Mines of the Anthracite Coal Regions of Pennsylvania, 1885*. Edwin K. Myers, 1886.
Pennsylvania Department of Public Education. *Safe Practices in Mining Anthracite*. Pennsylvania Department of Public Education, 1944.
"Pennsylvania Gas, Water Co. Observing 100th Anniversary." *Scrantonian*, 21 May 1967: 8.
Pennsylvania General Assembly. *Constitution of the Commonwealth of Pennsylvania*. https://www.legis.state.pa.us/cfdocs/legis/LI/consCheck.cfm?txtType=HTM&ttl=0. Accessed 26 Mar. 2021.
Pennsylvania Historical and Museum Commission. "Pennsylvania Historical Marker Search." Search term "first electric cars." http://www.phmc.state.pa.us/apps/historical-markers.html. Accessed 23 Mar. 2021.
Pennsylvania Independent Oil & Gas Association. "PA Oil and Gas." https://pioga.org/education/pa-oil-and-gas/. Accessed 23 Mar. 2021.
"People Flee from Homes." *Philadelphia Times*, 16 Dec. 1901: 2.
Perry, Daniel K. *"A Fine Substantial Piece of Masonry": Scranton's Historic Furnaces*. Pennsylvania Historical and Museum Commission, 1994.
"Perseverance Means Success." Advertisement. *Scranton Times*, 18 Nov. 1910: 16.
"Personal and Pertinent." *Scranton Times*, 14 Dec. 1906: 6.
Phillips, Susan. "Federal Public Health Report Highlights Contaminants in Dimock's Water." NPR. *StateImpact Pennsylvania*, 25 May 2016. https://stateimpact.npr.org

/pennsylvania/2016/05/25/federal-public-health-report-highlights-contaminants-in-dimocks-water/. Accessed 24 Mar. 2021.

Phillis, Michael, Matthew Daly, and John Flesher. "Biden Administration Weakens Water Protections after Supreme Court Curtails Federal Power." PBS. *News Hour*, 29 Aug. 2023. https://www.pbs.org/newshour/politics/biden-administration-weakens-water-protections-after-supreme-court-curtails-federal-power. Accessed 24 Oct. 2023.

"Phoebe Snow Replies to Admiral Evans." *New-York Tribune*, 19 Feb. 1907: 5.

"Phoebe Snow Symbolized Cleanliness." *Scranton Tribune*, 24 May 1983: 32.

Pinchot, Gifford. "Giant Power." *Survey Graphic*, 1 Mar. 1924: 561–562.

Pisani, Donald. *To Reclaim a Divided West: Water, Law, and Public Policy, 1848–1902*. University of New Mexico Press, 1992.

"Pittston." *Scranton Republican*, 24 July 1902: 8.

"Pittston." *Scranton Tribune*, 17 Aug. 1900: 2.

"Pittston City News." *Scranton Times*, 9 Aug. 1902: 7.

Plato. *Theaetetus*. Translated with an essay by Robin A. H. Waterfield. Penguin, 2004.

"Plenty of Water." *Wilkes-Barre Record*, 18 Sept. 1895: 1.

"Pollution Fight Is Planned." *Scrantonian*, 25 June 1944: 9.

Pontifical Academy of Sciences. "Final Statement of the Workshop on the Human Right to Water." https://www.pas.va/content/pas/en/events/2017/water/statement-water-eng.pdf. Accessed 7 May 2024.

Pope Francis. *On Care for Our Common Home* [*Laudato Si'*]. Encyclical letter. 2015.

Postel, Sandra. "Water and Agriculture." *Water in Crisis: A Guide to the World's Fresh Water Resources*, edited by Peter H. Gleick, 56–66. Oxford University Press, 1993.

"The Pot Holes Secured." *Scranton Weekly Republican*, 12 Oct. 1887: 3.

Powell, H. Benjamin. *Philadelphia's First Fuel Crisis: Jacob Cist and the Developing Market for Pennsylvania Anthracite*. Penn State University Press, 1978.

Powell, S. Robert. "Chronology: The Delaware and Hudson Canal Company and the Delaware and Hudson Company." 1985. Archives of the Lackawanna Historical Society.

———. *Delaware and Hudson Canal Company: Gravity Railroad. Waterpower*. Disc 6. DVD. Carbondale Historical Society, 2015.

"Power Service Proves Adequate for All Needs Here." *Wilkes-Barre Times Leader*, 12 Apr. 1926: 80.

PPL Corporation. "About Us." https://www.pplweb.com/who-we-are/about-us/. Accessed 14 Jan. 2021.

———. "Our Companies." https://www.pplweb.com/who-we-are/our-companies/. Accessed 14 Jan. 2021.

"Principal Kinds of Illuminating Gas." *Gas Industry* 12, no. 9 (Sept. 1912): 549–551.

Prochaska, Ernst. *Coal Washing*. McGraw-Hill, 1921.

"Proposed Water Works." *Scranton Republican*, 29 July 1889: 3.

"Public Baths for Scranton." *Scranton Tribune*, 11 Aug. 1898: 4.

"Public Meetings." *Wyoming Herald* 12 Feb. 1830: 3.

"Pumping a Million Tons of Coal from River-Beds." *Popular Science Monthly* 99 (Dec. 1921): 47.

"Purity of City Water." *Scranton Republican*, 10 Sept. 1895: 7.

"Raging Flood Leaves Wreck and Widespread Desolation in its Path." *Scranton Times*, 10 Oct. 1903: 1, 3.

"Railroad and Mine." *Wilkes-Barre Record*, 31 Oct. 1895: 7.

"Rates Are To [*sic*] High." *Scranton Republican*, 17 Dec. 1907: 3, 10.

Read, John. Letter to Thomas Meredith. 24 August 1820. Meredith Papers. Jenkins Manuscripts, box 13, 53/2. Archives of Lackawanna Historical Society.

Reese, Stuart O. "Outstanding Geologic Feature of Pennsylvania: Archbald Pothole, Lackawanna County." Trail of Geology 16-025.0. Pennsylvania Geological Society, 2016. https://elibrary.dcnr.pa.gov/PDFProvider.ashx?action=PDFStream&docID=1753466&chksum=&revision=0&docName=0032559&nativeExt=pdf&PromptToSave=False&Size=553296&ViewerMode=2&overlay=0. Accessed 9 May 2024.

Reisner, Marc. *Cadillac Desert: The American West and Its Disappearing Water.* Penguin, 1993. Repr., 1986.

"A Remarkable Man." In "Samuel Meredith" file. Archives of the Susquehanna County Historical Society, Montrose, PA.

Renewable Energy Aggregators. "Old Forge: Converting Contaminated Water into Clean Energy." 2021. https://reaggregators.com/projects/. Accessed 7 May 2024.

"Reservoirs at Scranton Gas Works Saved by Dynamite." *Philadelphia Inquirer*, 3 Mar. 1902: 2.

"A Retaining Wall." *Scranton Tribune*, 29 Mar. 1902: 12.

"Revolvers Fired at Pittston." *Scranton Republican*, 22 May 1902: 7.

Reynolds, Terry S. *Stronger Than a Hundred Men: A History of the Vertical Water Wheel.* Johns Hopkins University Press, 1983.

"Richmondale Hydro Plant Plans Are Moving Forward." *Forest City News*, 16 May 2018: 2.

"Richmondale Pumped Storage Hydroelectric Project, US." *Power Technology*, 10 Dec. 2021. https://www.power-technology.com/marketdata/richmondale-pumped-storage-hydroelectric-project-us/. Accessed 14 March 2023.

"A Riot at Pittston." *Scranton Republican*, 21 May 1902: 5.

"Rioting at Throop." *Scranton Republican*, 12 Aug. 1902: 5.

Ritter, Thomas. "Art. XVII—Notice of the Belmont Anthracite Mines, in Pennsylvania." *American Journal of Science* 12 (1827): 301–302.

"River Bed Has Changed." *Scranton Republican*, 28 Dec. 1899: 5.

"The Road of Anthracite." *Buffalo Evening News*, 11 Apr. 1903: 19.

Roberts, Ellis W. "A History of Land Subsidence and Its Consequences Caused by the Mining of Anthracite Coal in Luzerne County, Pennsylvania." Ph.D. diss., New York University, 1948.

Roberts, Peter. *Anthracite Coal Communities.* MacMillan, [1901] 1904.

Robertson, Craig G. "The Butler Water Tunnel Contamination in Pittston." PowerPoint presentation at the Significant Environmental Issues in Anthracite Country panel, Anthracite Mining History Conference, Marywood University, Scranton, PA, 18 Jan. 2019.

Robertson, Kirsty. "Oil Futures/Petrotextiles." In *Petrocultures: Oil, Politics, Culture*, edited by Sheena Wilson, Adam Carlson, and Imre Szeman, 242–263. McGill-Queen's University Press, 2017.

Rogers, Ed. "Wayne Will Get Economic Transfusion with Start of $30 Million Pipe Line Job." *Scranton Times*, 21 Apr. 1955: 3, 8.

Rojanasakul, Mira, Christopher Flavelle, Blacki Migliozzi, and Eli Murray. "America Is Using Up Its Groundwater Like There's No Tomorrow." *New York Times*, 28 Aug.

2023. https://www.nytimes.com/interactive/2023/08/28/climate/groundwater-drying-climate-change.html. Accessed 28 Oct. 2023.

Romm, Joseph. *Climate Change: What Everyone Needs to Know.* Oxford University Press, 2016.

Ronquillo, Romina. "Understanding Heat Exchangers." Thomas Publishing, 2020. https://www.thomasnet.com/articles/process-equipment/understanding-heat-exchangers/#register. Accessed 17 Sept. 2020.

Rosen, Julia. "Snowball Earth." *Scranton Times-Tribune*, 15 Jan. 2019: B7.

Rosler, Daniel. "Light at the End of the Borehole." *Scranton Times-Tribune*, 25 June 2018: A3.

"Rotogravure Section." *Scrantonian*, 30 August 1942: 89–92.

Roy, Arundhati. "The Greater Common Good." *Frontline*, 22 May 1999. https://web.cecs.pdx.edu/~sheard/course/Design&Society/Readings/Narmada/greatercommongood.pdf. Accessed 10 May 2024.

Rubinkam, Michael. "County, Public Left in Dark about Methane Leak." *Scranton Times-Tribune*, 25 Sept. 2017: A1, A8.

———. "Delaware River Panel Wants Ban on Fracking." *Scranton Times-Tribune*, 14 Sept. 2017: A6.

———. "Fracking Ban Gets Tougher." *Scranton Times-Tribune*, 26 Feb. 2021: A1, A7.

Rudzinski, Joe. "River Recollections: Throop." *History Bytes*, Mar.–Apr. 2020: 13–15.

Ruth, Philip. *Of Pulleys and Ropes and Gear: The Gravity Railroads of the Delaware and Hudson Canal Company and the Pennsylvania Coal Company.* Wayne County Historical Society, 1997.

Salzman, James. *Drinking Water: A History.* Overlook Duckworth, 2017.

Sanderson, Dorothy Hurlburt. *The Delaware & Hudson Canalway: Carrying Coals to Rondout.* 2nd ed. Rondout Valley Publishing, 1974.

"Sauquoit Mills to Shut Down." *Scranton Republican*, 29 May 1902: 5.

Sawyer, A. H. "Change House with Swimming Pools." *Engineering and Mining Journal* 98, no. 12 (Sept. 1914): 483–484.

Schechner, William. "Phoebe Snow's Friends Sadly Say Farewell." *Hackensack Record*, 28 Nov. 1966: A10.

Schmidt, Jeremy J. *Water: Abundance, Scarcity, and Security in the Age of Humanity.* New York University Press, 2017.

Schneider-Mayerson, Matthew. "'Just as in the Book?' The Influence of Literature on Readers' Awareness of Climate Justice and Perception of Climate Migrants." *ISLE: Interdisciplinary Studies in Literature and Environment* 27, no. 2 (Spring 2020): 337–364.

Scholz, Frank. "State Says Man Dumped Toxic Wastes under City." *Scranton Times*, 7 July 1983: 1, 20.

Schultz, Sally M., and Joan Hollister. "The Delaware and Hudson Canal Company: Forming, Financing, and Reporting on an Early 19th Century Corporation." *Accounting Historians Journal* 41, no. 2 (Dec. 2014): 111–152.

"Scranton Corp. Begins 66th Year." *Scrantonian*, 10 Mar. 1963: 15.

"Scranton Gas & Water Has Acquired Properties of Consolidated Company." *Scranton Truth*, 27 June 1905: 2.

"Scranton's Greatest Flood." *Scranton Republican*, 10 Oct. 1903: 5, 9.

"Scranton's Water Supply." *Scranton Times*, 23 Sept. 1895: 4.

"Scranton to Be Made a City of Lights by Combine." *Scranton Times*, 16 Mar. 1907: 1.

"Scranton Wants the Earth." *Scranton Republican*, 20 July 1889: 5.
Scripps Institution of Oceanography, University of California, San Diego. "The Keeling Curve." https://keelingcurve.ucsd.edu/. Accessed 24 Nov. 2023.
Sedlak, David. *Water 4.0: The Past, Present, and Future of the World's Most Vital Resource.* Yale University Press, 2014.
Sedlisky, Rick. "Huber: The End of an Era." In *Delaware and Hudson Canal Company Breakers*, edited by S. Robert Powell, 206. Carbondale Historical Society, 2017. https://archive.org/stream/01GravityRailroad1829Configuration/Delaware%20and%20Hudson%20Railroad/18%20Breakers_djvu.txt. Accessed 7 May 2024.
"Seek $305,000,000 for Flood Prevention." *Scranton Republican*, 26 Mar. 1936: 1.
Selig, E. T. "River Coal for Heating School Buildings." *Heating and Ventilating Magazine* 20, no. 4 (Apr. 1923): 43–44.
"Selling on Old Orders." *Wilkes-Barre Record*, 4 Nov. 1895: 8.
"Senate Added 12 Millions [*sic*] to Flood Bill." *Pittston Gazette*, 25 July 1941: 2.
Sengupta, Somini. "The Message from a Scorching 2018: We're Not Prepared for Global Warming." *Scranton Times-Tribune*, 10 Aug. 2018: B7.
"Serious Flood in Schooley Mine." *Wilkes-Barre Times*, 12 May 1899: 7.
"Seventeen Collieries Now in Operation." *Scranton Republican*, 29 Aug. 1902: 5.
"Seven Under Arrest." *Scranton Republican*, 28 June 1902: 1.
Shackel, Paul A. *Remembering Lattimer: Labor, Migration, and Race in Pennsylvania Anthracite Country.* University of Illinois Press, 2018.
"Shamokin's New Water Supply." *Carbondale Leader*, 2 Oct. 1895: 1.
Shaughnessy, Jim. *Delaware & Hudson: The History of an Important Railroad Whose Antecedent Was a Canal Network to Transport Coal.* Syracuse University Press, 1997.
Sheriff, Carol. "'Not the True Centennial': The Politics of Erie Canal Celebrations, 1917–1926." *New York History* 99, no. 3/4 (Summer/Fall 2018): 370–408.
"Sheriff's Sales." *Luzerne Union* [Wilkes-Barre], 22 Apr. 1863: 1.
Shi, Hui, et al. "Global Decline in Ocean Memory over the 21st Century." *Science Advances* 8, no. 18 (May 2022): n.p. https://www.science.org/doi/10.1126/sciadv.abm3468. Accessed 10 May 2022.
Siegle, Lucy. *To Die For: Is Fashion Wearing Out the World?* Fourth Estate, 2011.
"The Silk Industry in Scranton." *Scranton Republican*, 8 Dec. 1891: 8–9.
"Silk Mill Hands on Strike." *Scranton Tribune*, 25 Jan. 1901: 8.
"The Silk Mill Strike." *Scranton Tribune*, 12 Nov. 1900: 2.
"Silk Mills Strike." *Scranton Republican*, 4 Feb. 1901: 3.
"The Silk Workers' Wages." *Scranton Tribune*, 6 Feb. 1901: 2.
Silliman, Benjamin. "Notice of the Anthracite Region." *Register of Pennsylvania* 6, no. 5 (July 1830): 70–77.
Singleton, David. "Efforts to Save Wayne County Dam Underway." *Scranton Times-Tribune*, 18 Feb. 2019: A3.
———. "Historic Dam Gets Reprieve; No Demolition." *Scranton Times-Tribune*, 14 Oct. 2019: A3.
———. "Wayne County Wins Injunction against Historic Dam Demolition." *Scranton Times-Tribune*, 8 Sept. 2018: A4.
"Six Inches Rainfall." *Scranton Republican*, 6 Oct. 1902: 7.
Smil, Vaclav. *Energy: A Beginner's Guide.* 2nd ed. Oneworld Publishing, 2017.

Smith, Ernest G. "The Story of Anthracite's Creation." *Wilkes-Barre Times Leader*, 2 May 1942: 15.
"Source of the Lackawanna: Rival Companies Fighting to Secure It for a Water Supply." *Scranton Tribune*, 16 Dec. 1895: 3.
State, Ray. "The Truth behind the 'Pride of Newcastle.'" *Railroad History* 201 (Fall–Winter 2009): 71–85.
"State, County Officials to Seek a State Park." *Scranton Times*, 18 Mar. 1967: 3.
"Status at the Mines." *Scranton Republican*, 5 Nov. 1902: 5.
Steinberg, Theodore. *Nature Incorporated: Industrialization and the Waters of New England*. Cambridge University Press, 1991.
Stepenoff, Bonnie. *Their Fathers' Daughters: Silk Mill Workers in Northeastern Pennsylvania, 1880–1960*. Susquehanna University Press, 1999.
Stevenson, George E. *Reflections of an Anthracite Engineer*. Privately printed, 1931. HathiTrust. https://babel.hathitrust.org/cgi/pt?id=mdp.39015035043572&view=1up&seq=1. Accessed 8 May 2024.
Stewart, A. H. "Stream Pollution Control in Pennsylvania." *Sewage Works Journal* 17, no. 3 (May 1945): 586–593.
Stocker, Rhamanthus M. *Centennial History of Susquehanna County, Pennsylvania*. R. T. Peck, 1887.
"Stockport and Mount Pleasant Turnpike Road." *Susquehanna Democrat*, 27 June 1817: 4.
"Stockton Colliery Men Idle." *Pottsville Miners Journal*, 17 Sept. 1895: 4.
"The Story of Phoebe Snow." *Buffalo Times*, 13 Dec. 1903: 7.
Stranahan, Susan Q. *Susquehanna, River of Dreams*. Johns Hopkins University Press, 1995.
Strang, Veronica. *Water*. Reaktion Books, 2015.
"Streams Polluted with Mine Refuse." *Scranton Republican*, 6 Nov. 1902: 5.
"Strikers Gaining Ground." *Scranton Tribune*, 5 Jan. 1901: 2.
"Strike Situation Is Reviewed." *Scranton Tribune*, 29 May 1902: 1.
Sturluson, Snorri. *The Prose Edda: Tales from Norse Mythology*. Translated by Jean I. Young. University of California Press, 1954.
"A Supplement to the Act Entitled, 'An Act to Improve the Navigation of the River Lackawaxen.'" 1 Apr. 1825. In *Acts of the General Assembly of the Commonwealth of Pennsylvania*, 141–143. Mowry & Cameron, 1825.
"The Supreme Court." *Scranton Republican*, 7 Mar. 1893: 3.
"Switchmen to Act." *Scranton Republican*, 21 Sept. 1900: 3.
Taber, Thomas Townsend, and Thomas Townsend Taber III. *The Delaware, Lackawanna, & Western Railroad in the Twentieth Century, 1899–1960*. Vol. 1. Thomas T. Taber, 1980.
———. *The Delaware, Lackawanna, & Western Railroad in the Twentieth Century, 1899–1960*. Vol. 2. Thomas T. Taber, 1981.
Tabuchi, Hiroko, and Blacki Migliozzi. "'Monster Fracks' Are Getting Far Bigger. And Far Thirstier." *New York Times*, 26 Sept. 2023. https://www.nytimes.com/interactive/2023/09/25/ climate/fracking-oil-gas-wells-water.html. Accessed 28 Oct. 2023.
"Taking Coal from River." *Black Diamond* 55, no. 7 (14 Aug. 1915): 124.
Tarr, Joel A. "Toxic Legacy: The Environmental Impact of the Manufactured Gas Industry in the United States." *Technology and Culture* 55, no. 1 (Jan. 2014): 107–147.

Tarr, Joel A., and Terry F. Yosie. "Critical Decisions in Pittsburgh Water and Wastewater Treatment." In *Devastation and Renewal: An Environmental History of Pittsburgh and its Region*, edited by Joel A. Tarr, 64–88. University of Pittsburgh Press, 2003.
"Taylor." *Scranton Republican*, 18 July 1902: 2.
"Teams Wanted." *Wyoming Herald* [Wilkes-Barre], 20 Nov. 1829: 4.
"$10,000 a Year." *Scranton Tribune*, 28 July 1899: 7.
"Text of Opinion in Archbald Suit." *Scranton Times*, 9 Aug. 1902: 2.
"That Gigantic Scheme." *Wilkes-Barre Record*, 16 Mar. 1896: 2.
"That Slag in the River." *Scranton Republican*, 22 Dec. 1892: 5.
"Then & Now." *Scranton Sunday Times*, 5 Mar. 2017: A13.
"Then & Now." *Scranton Sunday Times*, 10 Feb. 2019: A13.
"They Are Happy Now." *Scranton Republican*, 1 May 1901: 7.
"They Won't Be Reinstated." *Scranton Tribune*, 30 May 1902: 7.
"Three Hundred Perish in Japanese Mine When Sea Floods Workings." *Pittston Gazette*, 13 Apr. 1915: 1.
"3,500 Lights in New Sign of Electric Co." *Scranton Republican*, 11 Oct. 1916: 3.
Throop, Benjamin. *A Half Century in Scranton*. Press of *Scranton Republican*, 1895.
"Time Extension Refused." *Scranton Tribune*, 18 Feb. 1896: 3.
"Today the Big Battle Begins." *Scranton Republican*, 17 Sept. 1900: 3.
"Today Will Decide." *Scranton Republican*, 2 June 1902: 1.
"To Flood the Mine." *Scranton Republican*, 14 July 1902: 7.
"Told in Brief." *Wilkes-Barre Record*, 30 June 1899: 16.
Tonge, James. "Modern Methods in Shaft Sinking." *Mines and Minerals*, May 1906: 467–468.
"Tools Lost in Mining Flood Worth $100,000." *Scranton Republican*, 26 Mar. 1936: 5.
"To Operate Mines." *Scranton Republican*, 9 June 1902: 5.
"To See Mr. Mitchell." *Scranton Republican*, 30 May 1902: 3.
Trails Conservation Corporation. *Upper Lackawanna Watershed Management Plan*. 2002. http://elibrary.dcnr.pa.gov/GetDocument?docId=1736855&DocName=RiverRegistry57_UpperLackawanna.pdf. Accessed 19 Mar. 2021.
Tramontana, Shirley. "Black Iron, White Lace." *St. Lawrence County Historical Society Association Quarterly* 38, no. 4 (Fall 1993): 3–14.
"The Trial Trip." *Scranton Republican*, 30 Nov. 1886: 3.
"Tried to Wreck a Train." *Scranton Republican*, 28 Sept. 1900: 1.
"Troops Now There." *Scranton Republican*, 20 Aug. 1902: 1
Trump, Donald J. "President Donald J. Trump's State of the Union Address." Prepared speech. 30 Jan. 2018. https://trumpwhitehouse.archives.gov/briefings-statements/president-donald-j-trumps-state-union-address/. Accessed 12 Mar. 2021.
Tufts University. "Miletus (Site)." Perseus 4.0. Perseus Digital Library. http://www.perseus.tufts.edu/hopper/artifact?name=Miletus&object=site. Accessed 13 Jan. 2021.
"Turbine Mine Pumps." *Colliery Engineer*, Dec. 1911: 287.
"Turbine Pumps in the Anthracite Coal Field." *Engineering and Mining Journal*, 14 Oct. 1905: 687–688.
Tvedt, Terje. "Why England and Not China and India? Water Systems and the History of the Industrial Revolution." *Journal of Global History* 5 (2010): 29–50.
Tvedt, Terje, and Terje Oestigaard. "A History of the Ideas of Water: Deconstructing Nature and Constructing Society." In *A History of Water: The Idea of Water from An-*

tiquity to Modern Times, vol. 4, edited by Terje Tvedt and Terje Oestigaard, 1–36. I.B. Tauris, 2010. https://www.researchgate.net/publication/287595904_A_History_of_Water_Series_2_Vol_1_Ideas_of_Water_from_Antiquity_to_Modern_Times. Accessed 13 Mar. 2021.

"12 Missing and 35 Rescued after River Breaks into Mine near Pittston." *Wilkes-Barre Record*, 23 Jan. 1959: 1, 2.

"28 Typhoid Cases in Hospital; Keller Says 'Not Alarmed.'" *Scranton Times*, 27 Aug. 1909: 1.

"A Two Days' Rain Needed." *Carbondale Leader*, 5 Oct. 1895: 1.

"Two Mine Horrors." *Lancaster Daily New Era*, 5 Feb. 1891: 1.

"Two Serious Riots." *Scranton Republican*, 23 Oct. 1900: 1.

"The Typhoid Situation." *Wilkes-Barre Record*, 9 Sept. 1905: 5.

"Uniondale." *Scranton Tribune*, 26 June 1897: 12.

United Nations. "International Decade for Action on Water for Sustainable Development, 2018–2028." https://www.un.org/en/events/waterdecade/. Accessed 13 Mar. 2021.

United States Tariff Commission. *Report on Dyes and Related Coal-Tar Chemicals 1918*. GPO, 1919.

"Unlimited in Quantity; Unexcelled in Quality." *Scranton Times*, 12 Oct. 1926: 38.

U.S. Department of Energy. *The Water-Energy Nexus: Challenges and Opportunities*. 2014. https://www.energy.gov/sites/prod/files/2014/07/f17/Water%20Energy%20Nexus%20Ful l%20Report%20July%202014.pdf. Accessed 22 Mar. 2021.

"U.S. Drinking Water Contamination with 'Forever Chemicals' Far Worse than Scientists Thought." *The Guardian*, 22 Jan. 2020. https://www.theguardian.com/environment/2020/jan/22/us-drinking-water-contamination-forever-chemicals-pfas. Accessed 21 July 2020.

U.S. Geological Survey. "The Water in You: Water and the Human Body." Water Science School. 22 May 2019. https://www.usgs.gov/special-topic/water-science-school/science/water-you-water-and-human-body?qt-science_center_objects=0#qt-science_center_objects. Accessed 27 Mar. 2021.

———. "Watersheds and Drainage Basins." Water Science School. 8 June 2019. https://www.usgs.gov/special-topic/water-science-school/science/watersheds-and-drainage-basins?qt-science_center_objects=0#qt-science_center_objects. Accessed 17 Jan. 2021.

U.S. Global Change Research Program. "Key Message #7: Changes to Water Demand and Use." In *Highlights: Water Supply. Our Changing Climate: Water Resources*. National Climate Assessment. http://nca2014.globalchange.gov/report/sectors/water. Accessed 22 Mar. 2021.

"Valley Is Swept by a Terrific Storm." *Scranton Republican*, 16 Dec. 1901: 3, 4.

"Very Exciting Day." *Scranton Republican*, 24 Oct. 1900: 1.

"Views on the Water Plant." *Scranton Tribune*, 26 Dec. 1901: 3.

Voss-Hubbard, Anke. "Hugh Moore and the Selling of the Paper Cup: A History of the Dixie Cup Company, 1907–1957." *Canal History and Technology Proceedings* 15 (Mar. 1996): 83–101.

Wagner, Van. *The Billy Marks: Paddlewheel Coal Diggers on the Susquehanna River*. Documentary film, 2012. http://www.vanwagnermusic.com/coaldredging.html. Accessed 8 May 2024.

———. "The Hard Coal Navy of the Susquehanna River." July 2012. http://www.vanwagnermusic.com/coaldredging.html. Accessed 8 May 2024.

Walcott, Charles D. *Twenty-Fourth Annual Report of the Director of the United States Geological Survey to the Secretary of the Interior, 1902–3*. U.S. Geological Survey. GPO, 1903. https://doi.org/10.3133/ar24.

Wallace, Anthony F. C. *St. Clair: A Nineteenth-Century Coal Town's Experience with a Disaster-Prone Industry*. Knopf, 1987.

Wallace, David Foster. Commencement address, Kenyon College, 2005. "This is Water," available at archive of *Kenyon College Alumni Bulletin*. http://bulletin-archive.kenyon.edu/x4280.html. Accessed 8 May 2024.

Walsh, Owen. "Fight over Fracking." *River Reporter*, 6 Jan. 2021. https://riverreporter.com/stories/fight-over-fracking,41247. 24 Mar. 2021.

Walster Corporation to Doyle & Roth Manufacturing Company. Land deed, Fell Township. 23 July 1974. Bk. 832, pp. 152–159. Recorder of Deeds Office, Lackawanna County, PA.

"Washeries to be Closed." *Scranton Republican*, 8 Sept. 1900: 5.

"Washery Consumed by a Mysterious Fire." *Scranton Republican*, 28 Aug. 1902: 5.

"Washery Stopped." *Scranton Republican*, 1 July 1902: 6.

"Washery Was Shut Down." *Scranton Republican*, 1 Nov. 1900: 5.

"Washing Anthracite." *Wilkes-Barre Record*, 19 Dec. 1904: 15.

"Water Co. Gets Charter." *Wilkes-Barre Times*, 11 Sept. 1903: 7.

"Water Companies—Lessors." In *Second Industrial Directory of Pennsylvania*, 768–775. William Stanley Ray, 1916.

"Water Company's Scheme." *Scranton Tribune*, 16 Oct. 1895: 3.

"Water Famine." *Carbondale Leader*, 13 Sept. 1895: 4.

"The Water Famine." *Scranton Republican*, 27 Sept. 1895: 1.

"Water Famine in the West." *Scranton Republican*, 12 Oct. 1895: 2.

"The Water Not Good." *Scranton Republican*, 7 Aug. 1889: 5.

"Water Rates Again." *Scranton Republican*, 19 Dec. 1899: 5.

"Water Recedes, Damage Great." *Scranton Times*, 12 Oct. 1903: 3.

"Water Still Coming In." *Wilkes-Barre Record*, 16 May 1899: 8.

Water Supply Commission of Pennsylvania. *Report of the Water Supply Commission of Pennsylvania, 1908*. C. E. Aughinbaugh, 1910.

———. *Water Resources Inventory Report*. Act of July 25, 1913. Part 10, "Culm in the Streams of the Anthracite Region." Water Supply Commission of Pennsylvania, 1916. https://books.google.com/books?id=klhBAQAAMAAJ&pg=RA2-PA20&lpg=RA2-PA20&dq=%E2%80%9Cculm+and+mine+water+from+67+collieries+and+numerous+m ine+openings%E2%80%9D&source=bl&ots=1JetoJPEPS&sig=ACfU3U0Jno0396nXCn QOfN8g4pCUsHUi4A&hl=en&sa=X&ved=2ahUKEwjO3vWgm7DvAhXI1VkKHWuI BCUQ6AEwAXoECAEQAw#v=onepage&q=part%20x&f=false. Accessed 14 Mar. 2021.

"The Water Supply of Scranton." *Scranton Republican*, 14 Dec. 1895: 6.

"Water Thwarts Mayor from Entering Mine." *Scranton Times*, 15 Mar. 1920: 1, 2.

Watts, Jonathan. "Concrete: The Most Destructive Material on Earth." *The Guardian*, 25 Feb. 2019. https://www.theguardian.com/cities/2019/feb/25/concrete-the-most-destructive-material-on-earth. Accessed 24 Mar. 2021.

"Wayne County." *Register of Pennsylvania* 3, no. 9 (Feb. 1829): 135–139.

Weaver, Karol K. *Medical Caregiving and Identity in Pennsylvania's Anthracite Region, 1880–2000*. Penn State University Press, 2011.

Webber, Michael E. *Thirst for Power: Energy, Water, and Human Survival*. Yale University Press, 2016.

Weissert, Will. "Big Oil Seeking Protection from Climate Change from Government." *Scranton Times-Tribune*, 23 Aug. 2018: B11.

West Virginia v. EPA, 20 U.S. 1530 (2022). https://www.supremecourt.gov/opinions/21pdf/20-1530_n758.pdf. Accessed 10 May 2024.

Wharton School of the University of Pennsylvania. "Ebb without Flow: Water May Be the New Oil in a Thirsty Global Economy." *Knowledge at Wharton*, 1 Oct. 2008. https://knowledge.wharton.upenn.edu/article/ebb-without-flow-water-may-be-the-new-oil-in-a-thirsty-global-economy/. Accessed 19 Mar. 2021.

"What Some People Say." *Scranton Republican*, 13 Jan. 1897: 11.

"Where Responsibility for Flood Damages Lies." *Scranton Times*, 12 Oct. 1903: 4.

White, Gilbert F. Foreword to *Water in Crisis: A Guide to the World's Fresh Water Resources*, edited by Peter H. Gleick, n. p. Oxford University Press, 1993.

White, I. C. *Second Geological Survey of Pennsylvania: Report of Progress / The Geology of Susquehanna County and Wayne County*. Board of Commissioners for the Second Geological Survey, 1881.

White, Joseph Hill. *Miner's Wash and Change Houses*. GPO, 1915.

"Widen River Channel, and Prevent Floods." *Scranton Times*, 16 Oct. 1903: 5, 11.

Wilber, Tom. *Under the Surface: Fracking, Fortunes, and the Fate of the Marcellus Shale*. Cornell University Press, 2015. Repr., 2012.

"Will Made by John Jermyn." *Scranton Tribune*, 19 July 1902: 6.

"Will Not Leave Work." *Scranton Republican*, 29 May 1902: 5.

"Will Not Strike." *Scranton Republican*, 27 May 1902: 7.

"Wilsonville Dam." *Milford Dispatch*, 7 July 1910: 1.

Wind, Kyle. "Plant Won't Discharge Wastewater into Creek." *Scranton Times-Tribune*, 8 July 2017: A1, A10.

Wolensky, Kenneth C., Nicole H. Wolensky, and Robert P. Wolensky. *Fighting for the Union Label: The Women's Garment Industry and the ILGWU in Pennsylvania*. Penn State University Press, 2002.

Wolensky, Robert P. *Sewn in Coal Country: An Oral History of the Ladies' Garment Industry in Northeastern Pennsylvania, 1945–1995*. Penn State University Press, 2020.

Wolensky, Robert P., and William A. Hastie, Sr. *Anthracite Labor Wars: Tenancy, Italians, and Organized Crime in the Northern Coalfield of Northeastern Pennsylvania, 1895–1959*. Canal History and Technology Press, 2013.

Wolensky, Robert P., Kenneth C. Wolensky, and Nicole H. Wolensky. *The Knox Mine Disaster, January 22, 1959: The Final Years of the Northern Anthracite Industry and the Effort to Rebuild a Regional Economy*. Pennsylvania Historical and Museum Commission, 1999.

Wordsworth, William. *The Prelude, 1799, 1805, 1850*, edited by Jonathan Wordsworth, M. H. Abrams, and Stephen Gill. W. W. Norton, 1979.

"Work at Von Storch." *Scranton Republican*, 22 Aug. 1902: 3.

"Working Anthracite Culm Piles." *Colliery Engineer* 33, no. 10 (May 1913): 569–571.

"Workmen Unearth Old Waterwheel." *Carbondale Leader*, 29 Jan. 1902: 5.

World Health Organization. "1 in 3 People Globally Do Not Have Access to Safe Drinking Water—UNICEF, WHO." 18 June 2019. https://www.who.int/news

/item/18-06-2019-1-in-3-people-globally-do-not-have-access-to-safe-drinking-water -unicef-who. Accessed 14 July 2020.

World Wildlife Fund. "Executive Summary." In *Living Planet Report 2020: Bending the Curve of Biodiversity Loss*, edited by R.E.A. Almond, M. Grooten, and T. Petersen, 6–8. World Wildlife Fund, 2020. https://c402277.ssl.cf1.rackcdn.com/publications /1371/files/original/ENGLISH-FULL.pdf?1599693362. Accessed 13 Jan. 2021.

Worster, Donald. *Rivers of Empire: Water, Aridity, and the Growth of the American West*. Pantheon, 1986.

"Would Seize All Anthracite." *Wilkes-Barre Semi-Weekly Record*, 8 Feb. 1907: 4.

Yaeger, Patricia. "Introduction: Dreaming of Infrastructure." *PMLA* 122, no. 1 (Jan. 2007): 9–26.

Yewlett, Hilary Lloyd. "Early Modern Migration from the Mid-Wales County of Radnorshire to Southeastern Pennsylvania, with Special Reference to Three Meredith Families." *Pennsylvania History* 79, no. 1 (2012): 1–32.

Young, Maryann Spellman. "Protect Lackawanna River's Natural Integrity." *Scranton Times-Tribune*, 4 June 2017: C2.

Zaitz, Anton. "Recollections of a Slovenian Miner." Part 6. 1940. Translated by Joe Drasler. *Forest City News*, 1 Oct. 2020: 10.

Index

Italicized numbers refer to illustrations.

Bill Conlogue is Professor of English at Marywood University and author of *Undermined in Coal Country: On the Measures in a Working Land, Here and There: Reading Pennsylvania's Working Landscapes,* and *Working the Garden: American Writers and the Industrialization of Agriculture.*